Soils of Malaysia

Soils of Malaysia

Edited by
Muhammad Aqeel Ashraf
Radziah Othman
Che Fauziah Ishak

CRC Press
Taylor & Francis Group
Boca Raton London New York

CRC Press is an imprint of the
Taylor & Francis Group, an **informa** business

CRC Press
Taylor & Francis Group
6000 Broken Sound Parkway NW, Suite 300
Boca Raton, FL 33487-2742

First issued in paperback 2021

© 2018 by Taylor & Francis Group, LLC
CRC Press is an imprint of Taylor & Francis Group, an Informa business

No claim to original U.S. Government works

ISBN-13: 978-1-03-209635-3 (pbk)
ISBN-13: 978-1-138-19769-5 (hbk)

Library of Congress Cataloging-in-Publication Data

Names: Ashraf, Muhammad Aqeel, editor.
Title: Soils of Malaysia / [edited by] Muhammad Aqeel Ashraf, Radziah Othman, and Che Fauziah Ishak.
Description: Boca Raton : Taylor & Francis, 2017.
Identifiers: LCCN 2017013875 | ISBN 9781138197695 (978-1-138-19769-5)
Subjects: LCSH: Soils--Malaysia.
Classification: LCC S599.6.M4 S655 2017 | DDC 631.4/9595--dc23
LC record available at https://lccn.loc.gov/2017013875

Visit the Taylor & Francis Web site at
http://www.taylorandfrancis.com

and the CRC Press Web site at
http://www.crcpress.com

Contents

Editors

Dr. Muhammad Aqeel Ashraf is working as a professor at the Department of Environmental Sciences and Engineering, School of Environmental Studies, China University of Geosciences, Beijing, China. He earned his PhD from the University of Malaya (UM), Kuala Lumpur, Malaysia. His research work focuses on environmental geochemistry especially soil and water science. He has published many books, book chapters, and research articles in this field.

Dr. Radziah Othman obtained her BSc degree (Microbiology) from Indiana University, Bloomington, Indiana, master's degree (Soil Science) from the North Carolina State University, Raleigh, North Carolina, and her PhD (Soil Microbiology) from Universiti Putra Malaysia, Selangor, Malaysia. She is an associate professor and currently the Head of Department of Land Management in the Faculty of Agriculture, Universiti Putra Malaysia. She specializes in the beneficial microorganisms for agriculture particularly in nitrogen-fixing bacteria, phosphate solubilizing bacteria, and mycorrhizae. She has published several scientific articles in cited journals and review articles in book chapters. She has developed biofertilizer for rice cultivation in acid sulfate soil and is now venturing into utilizing salt tolerant bacteria for alleviating salt stress in rice grown in salt affected areas.

Dr. Che Fauziah Ishak is a professor at the Department of Land Management, Faculty of Agriculture, Universiti Putra Malaysia, Selangor, Malaysia. She obtained her bachelor's and master's degree in chemistry from the University of Iowa, Iowa and PhD in agronomy from the University of Athens, Georgia. Her field of expertise is in organic matter and industrial biosolids (biochar, paper mill, and sewage sludge) recycling on agricultural land to improve soil fertility for crop improvement, with special emphasis on heavy metals uptake by plants, which is becoming an important issue in food safety and is a part of the food security agenda. Her research also includes management of metal-contaminated soils through in situ immobilization technique using industrial by-products such as red gypsum, coal fly ash, and water treatment residuals.

Contributors

Rosazlin Abdullah
Institute of Biological Sciences
University of Malaya
Kuala Lumpur, Malaysia

Muhammad Aqeel Ashraf
Department of Environmental Science
  and Engineering
School of Environmental Studies
China University of Geosciences
Wuhan, People's Republic of China

Md. Maniruzzaman A. Aziz
Department of Geotechnics and
  Transportation
Universiti Teknologi Malaysia
Johor, Malaysia

Samavia Batool
Economic Growth Unit
Sustainable Development Policy
  Institute
Islamabad, Pakistan

Wan Noordin Daud
Crop Science, Faculty of Agriculture
Universiti Putra Malaysia
Selangor, Malaysia

B.M. Firuza
Department of Geography
University of Malaya
Kuala Lumpur, Malaysia

Che Fauziah Ishak
Land Management, Faculty of
  Agriculture
Universiti Putra Malaysia
Selangor, Malaysia

Roslan Bin Ismail
Land Management, Faculty of
  Agriculture
Universiti Putra Malaysia
Selangor, Malaysia

Hawa Z.E. Jaafar
Crop Science, Faculty of Agriculture
Universiti Putra Malaysia
Selangor, Malaysia

Habibah Jamil
Environmental & Natural Resource
  Sciences
Universiti Kebangsaan Malaysia
Selangor, Malaysia

Aysha Masood Khan
National Center for Bioinformatics
Quaid-i-Azam University
Islamabad, Pakistan

Aminaton Marto
Department of Civil Engineering
Universiti Teknologi Malaysia
Johor, Malaysia

Nur Zurairahetty Mohd Yunus
Department of Geotechnics and
  Transportation
Universiti Teknologi Malaysia
Johor, Malaysia

Safiah Yusmah Mohd Yusoff
Department of Geography
University of Malaya
Kuala Lumpur, Malaysia

Shafar Jefri Mokhtar
Crop Science, Faculty of Agriculture
Universiti Putra Malaysia
Selangor, Malaysia

Rizwana Naureen
Department of Geology
University of Malaya
Kuala Lumpur, Malaysia

Radziah Othman
Land Management, Faculty of
   Agriculture
Universiti Putra Malaysia
Selangor, Malaysia

Qurban Ali Panhwar
Soil and Environmental Sciences
   Division
Nuclear Institute of Agriculture
Sindh, Pakistan

Zaharah A. Rahman
Land Management, Faculty of
   Agriculture
Universiti Putra Malaysia
Selangor, Malaysia

Amin Soltangheisi
Land Management, Faculty of
   Agriculture
Universiti Putra Malaysia
Selangor, Malaysia

Christopher Teh Boon Sung
Land Management, Faculty of
   Agriculture
Universiti Putra Malaysia
Selangor, Malaysia

M.B. Yusuf
Department of Geography
Taraba State University
Jalingo, Nigeria

1 Climate, Ecosystem, Flora, and Fauna

Hawa Z.E. Jaafar and Muhammad Aqeel Ashraf

CONTENTS

INTRODUCTION

Malaysia is located in the southeast of the Asian continent, a region referred to as Southeast Asia with coordinates of 2°30′N 112°30′E. It covers a total area of 329,847 km² (127,355 mi²), supporting over 28 millions of inhabitants in the country (2012). Malaysia land area consists of two distinct parts: Peninsular Malaysia (132,090 km²) to the west and East Malaysia (198,847 km²) to the east. These two

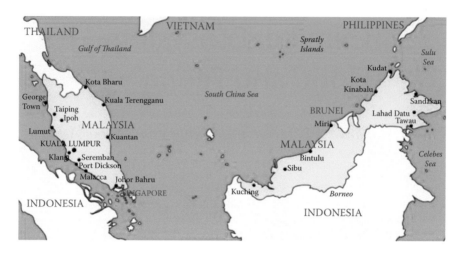

FIGURE 1.1   Political and administrative map of Malaysia. (From http://www.nationsonline. org/oneworld/map/malaysia_map.htm, retrieved November 10, 2014.)

parts of Malaysia occupying respective 39.7% and 60.3% of the country's land area are separated from each other by the South China Sea. They largely share a similar landscape in that both West (Peninsular) and East Malaysia feature coastal plains rising to hills and mountains (CIA 2006).

The former is located between Thailand in the north, Singapore in the south, and east of the Indonesian island of Sumatra. East Malaysia consisting of two provinces of Sabah and Sarawak is located on the island of Borneo and shares borders with Indonesia (Kalimantan) and Brunei as given in map of Malaysia in Figure 1.1. The highest point in Malaysia is Mount Kinabalu in Sabah; the longest river is Rajang River in Sarawak; and the largest lake is Bera Lake in Pahang, Semenanjung Malaysia. Of the total land area, 0.37% is made up of water bodies such as lakes, rivers, or other internal waters.

Malaysia has a total coastline of 4675 km: Peninsular Malaysia has 2068 km, whereas East Malaysia has 2607 km of coastline (CIA 2006). Peninsular Malaysia is mountainous, with more than half of it over 150 m above sea level. East Malaysia, on the other hand, is divided between coastal regions, hills and valleys, and a mountainous interior with only two major cities: Kuching and Kota Kinabalu. Much of southern Sarawak is coastal lowlands, which shifts to a series of plateaus going north, ending in the mountainous regions of Sabah (Marshall Cavendish Corporation 2008).

Malaysia, located near the equator, generally has climate characteristic features of uniform high temperature, high humidity, and copious rainfall throughout the year, although the geography of Malaysia may spell variations in the microclimate of the regions in the country. Winds are generally light. Situated in the equatorial doldrums area, it is extremely rare to have a full day with a completely clear sky even during periods of severe drought. On the other hand, it is also rare to have a stretch of a few days with completely no sunshine except during the northeast monsoon seasons.

THE CLIMATE OF MALAYSIA

Climate is a measure of the average pattern of variation in temperature, sunshine and solar radiation, precipitation, humidity, evaporation, wind, atmospheric pressure, atmospheric particle count, and other meteorological variables in a given region over long periods of time. A region's climate is generated by the climate system, which has five components: atmosphere, hydrosphere, cryosphere, lithosphere, and biosphere (Anon 2014a). The climate of a location is affected by its latitude, terrain, and altitude, as well as nearby water bodies and their currents. Climates can be classified according to the average and the typical ranges of different variables, most commonly temperature and precipitation.

Malaysia is categorized as equatorial with Semenanjung Malaysia that lies just above the equator and Borneo just below it, and being hot and humid throughout the year. The average temperature is 27°C (Saw 2007) with an average rainfall of 2500 mm a year (Anon 2008). In the dry season, Malaysia is exposed to the El Niño effect, which further reduces rainfall. Climate change is likely to have a significant effect on Malaysia, increasing sea levels and rainfall, increasing flooding risks, and leading to large droughts (Marshall Cavendish Corporation 2008).

TEMPERATURE

Malaysia, located in the equatorial region, has a tropical rainforest climate experiencing warm and humid condition throughout the year. In general, the climate of Malaysia can be described as typical tropical climate having a uniform temperature. The variations in the climate of Malaysia are dictated by the states and provinces that occupy definite climatological zones of Malaysia. Local climates are affected by the presence of mountain ranges throughout Malaysia. The climate, thus, can be divided into that of the highlands, the lowlands, and coastal regions. Temperature of the coastal plains averages at 28°C, the inland and mountain areas average at 26°C, and the higher mountain regions average at 23°C.

Malaysia essentially observes tropical weather, without extremely high temperatures. Humidity however is a common feature; nights in Malaysia are fairly cool. Throughout the year, the temperature ranges from 20°C to 30°C on an average. The coasts have a sunny climate, with temperatures ranging between 23°C and 32°C and rainfall ranging from 100 to 300 mm a month. The lowlands have a similar temperature but follow a more distinctive rainfall pattern and show very high humidity levels. The highlands, on the other hand, are cooler and wetter and display a greater temperature variation. A large amount of cloud cover is present over the highlands, with humidity levels that do not fall below 75% (Marshall Cavendish Corporation 2008).

The highest temperature recorded in 2012 at Subang station was 36.6°C compared to higher value recorded in Chuping, Perlis on April 9, 1998 at 40.1°C. The variation in the highest temperature could be due to variations in localized scenario. Meanwhile, the lowest temperature still remained at Cameron Highlands station except that the lowest temperature recorded in 2013 (13.0°C) had a nearly two-fold increase from that registered on February 1, 1978 (7.8°C) (Anon 2013; Malaysian Meteorological Department 2008).

SUNSHINE AND SOLAR RADIATION

Being a maritime country close to the equator, Malaysia naturally has abundant sunshine and, thus, solar radiation. The cloud cover cuts off a substantial amount of sunshine and, thus, solar radiation. On the average, Malaysia receives about 6 hours of sunshine per day. There are, however, seasonal and spatial variations in the amount of sunshine received. Alor Setar and Kota Bharu receive about 7 hours of sunshine per day, whereas Kuching and Kota Kinabalu receive only about 5 hours of sunshine on the average.

On the extreme, Kuching receives only an average of 3.7 hours per day in the month of January. On the other extreme, Alor Setar receives a maximum of 8.7 hours per day on the average in the same month. Solar radiation is closely related to the sunshine duration. Its seasonal and spatial variations are, therefore, very much the same as in the case of sunshine (Anon 2013).

RAINFALL DISTRIBUTION

Local climates are affected by the presence of mountain ranges throughout Malaysia. The seasonal wind flow patterns coupled with the local topographic features determine the rainfall distribution patterns over the country. During the northeast monsoon season, the exposed areas such as the east coast of Peninsular Malaysia, Western Sarawak, and the northeast coast of Sabah experience heavy rain spells. On the other hand, inland areas or areas that are sheltered by mountain ranges are relatively free from its influence.

On January 6, 1967, Kota Bharu in Kelantan recorded the highest rainfall in a day at 608 mm. The highest rainfall recorded in a year was 5687 mm at Sandakan, Sabah in 2006, and Sarawak registered an average rainfall of 4128 mm with 247 days of rain a year. Meanwhile, the lowest rainfall recorded in a year was 1151 mm at Tawau, Sabah in 1997. The driest place in Peninsular Malaysia is in Chuping, Perlis with an average rainfall of only 1746 mm a year (Malaysian Meteorological Department 2008).

In 2013, the wettest place in Malaysia is Gunung Gagau in Kelantan with a total annual rainfall of 5072 mm followed closely by Kg Besut, Setiu (4679 mm) and Calok Setiu (4059 mm), Terengganu as given in Table 1.1. In East Malaysia, Kapit and Kuching, Sarawak recorded the highest total annual rainfall of 4174.7 and 4167.2 mm, respectively. In 2012, the yearly highest and lowest rainfall were recorded at Bintulu station (3936.2 mm) and Sitiawan station (1833.8 mm), respectively. Total annual rainfall for 34 selected meteorological stations decreased from 101.4 thousand mm in 2011 to 91.7 thousand mm in 2012. Similarly, during the first 6 months of 2012 and 2013, total rainfall also decreased from 39.7 thousand mm to 37.9 thousand mm according to seasons (MMD 2014).

In Peninsular Malaysia, the highest number of rainy days was recorded in Cameron Highlands with 239 days in 2012. Meanwhile, Kota Bharu recorded the lowest number of rainy days with 171 days. In 2012, for Sabah, the lowest number of rainy days was recorded in Kota Kinabalu at 195 days, whereas the highest number of rainy days was recorded in Sandakan at 217 days. In Sarawak, the lowest number

TABLE 1.1
Malaysia—Projected and Actual Annual Rainfall Statistics for 2013 at Selected Locations

Number	Location (City/Town/Village)	State	Projected Total Average Rainfall (January–December 2013)		Actual Rainfall (January 1– December 31, 2013)	
			mm	in.	mm	in.
1	Batu 13 Jeli	Kelantan	6468.90	254.68	3810.50	150.02
2	Gunung Gagau	Kelantan	6207.00	244.40	5072.00	199.70
3	Pahang	Cherating	4533.00	178.46	3439.80	135.43
4	Kuching	Sarawak	4485.90	176.61	4167.20	164.06
5	Chalok Setiu	Terengganu	4475.10	176.19	4059.00	159.80
6	Kg Besut, Setiu	Terengganu	4438.60	174.75	4679.00	184.20
7	Bukit Larut, Taiping	Perak	4173.90	164.33	3616.00	142.40
8	Kapit	Sarawak	4126.40	162.46	4174.70	164.36
9	Bintulu	Sarawak	4091.70	161.09	3987.50	156.99
10	Gong Kedak	Terengganu	3908.90	153.89	3274.80	128.93
11	Sibu	Sarawak	3741.50	147.30	2774.80	109.24
12	Limbang	Sarawak	3661.40	144.15	3010.10	118.51
13	Sandakan	Sabah	3594.70	141.52	2677.80	105.43
14	Sri Aman	Sarawak	3543.60	139.51	2797.50	110.14
15	Labuan	Labuan	3364.70	132.47	3119.80	122.83
16	Petaling Jaya	Selangor	3288.70	129.48	3540.00	130.00
17	Kuantan	Pahang	3156.30	124.26	3572.90	140.67
18	Cameron Highlands	Pahang	2953.50	116.28	2827.20	111.31
19	Miri	Sarawak	2917.60	114.87	3168.00	124.70
20	Kudat	Sabah	2913.10	114.69	2039.60	80.30
21	Subang Jaya	Selangor	2900.70	114.20	2964.50	116.71
22	Ipoh	Perak	2881.60	113.45	2703.80	106.45
23	Kota Kinabalu	Sabah	2843.60	111.95	3169.60	124.79
24	Kuala Terengganu	Terengganu	2833.70	111.56	2945.50	115.96
25	Kuala Krai	Kelantan	2703.80	106.45	3168.80	124.76
26	Kota Bharu	Kelantan	2679.80	105.5	2226.20	87.65
27	Langkawi	Kedah	2533.90	99.76	2665.40	104.94
28	Mersing	Johor	2528.10	99.53	2905.20	114.38
29	Senai	Johor	2457.10	96.74	2876.40	113.24
30	Muadzam Shah	Pahang	2367.70	93.22	2887.30	113.67
31	Alor	Kedah	2329.00	91.69	1840.80	72.47
32	Bayan Lepas	Penang	2316.50	91.20	2357.00	92.80
33	Butterworth	Penang	2240.10	88.19	2767.90	108.97
34	Batu Embun, Jerantut	Pahang	2219.10	87.37	2581.10	101.62
35	Kluang	Johor	2137.00	84.13	2018.80	79.48

(Continued)

TABLE 1.1 (*Continued*)
Malaysia—Projected and Actual Annual Rainfall Statistics for 2013 at Selected Locations

Number	Location (City/Town/Village)	State	Projected Total Average Rainfall (January–December 2013)		Actual Rainfall (January 1–December 31, 2013)	
			mm	in.	mm	in.
36	Batu Pahat	Johor	2112.80	83.18	2031.80	79.99
37	Sepang KLIA	Selangor	2029.30	79.89	1637.60	64.47
38	Chuping	Perlis	2023.40	79.66	1735.60	68.33
39	Temerloh	Pahang	2016.60	79.39	2410.20	94.89
40	Lubuk Merbau	Perak	1915.90	75.43	1519.00	59.80
41	Sitiawan	Perak	1902.00	74.88	1954.30	76.94
42	Malacca City	Malacca	1884.20	74.18	1434.50	56.48

Source: http://en.wikipedia.org/wiki/Geography_of_Malaysia, retrieved November 14, 2014.

of rainy days was recorded in Miri at 203 days, whereas the highest number of rainy days was recorded in Kuching at 247 days (MMD 2014).

In 2013, the driest areas were registered in Malacca City, Malacca; Lubuk Merbau, Perak; Sepang Kuala Lumpur International Airport (KLIA), Selangor; and Chuping, Perlis, all recording between 143.4 and 173.5 mm thousands of total annual rainfall (Anon 2013). The projected and actual annual rainfall statistics in Malaysia for 2013 is best described in Table 1.1 (MMD 2014).

SEASONAL RAINFALL VARIATION IN PENINSULAR MALAYSIA

The seasonal variation of rainfall in Peninsular Malaysia is of three main types (MMD 2014):

1. Over the east coast states, November, December, and January are the months with maximum rainfall, whereas June and July are the driest months in most districts.
2. Over the rest of the Peninsula with the exception of the southwest coastal area, the monthly rainfall pattern shows two periods of maximum rainfall separated by two periods of minimum rainfall. The primary maximum generally occurs in October–November, whereas the secondary maximum generally occurs in April–May. Over the northwestern region, the primary minimum occurs in January–February with the secondary minimum in June–July, whereas elsewhere the primary minimum occurs in June–July with the secondary minimum in February.
3. The rainfall pattern over the southwest coastal area is much affected by early morning *Sumatras* from May to August with the result that the double

maxima and minima pattern is no longer distinguishable. October and November are the months with maximum rainfalls, and February is the month with the minimum rainfall. The March–April–May maximum and the June–July minimum rainfalls are absent or indistinct.

SEASONAL RAINFALL VARIATION IN SABAH AND SARAWAK

The seasonal variation of rainfall in Sabah and Sarawak can be divided into five main types (MMD 2014):

1. The coastal areas of Sarawak and northeast Sabah experience a rainfall regime of one maximum and one minimum. Although the maximum rainfall occurs during January in both areas, the occurrence of the minimum rainfall differs. In the coastal areas of Sarawak, the minimum rainfall occurs in June or July, whereas in the northeast coastal areas of Sabah, it occurs in April. Under this regime, much of the rainfall is received during the northeast monsoon months of December to March. In fact, it accounts for more than half of the annual rainfall received on the western part of Sarawak.
2. Inland areas of Sarawak generally experience quite evenly distributed annual rainfall. Nevertheless, slightly less rainfall is received during the period June to August, which corresponds to the occurrence of prevailing southwesterly winds. It must be pointed out that the highest annual rainfall area in Malaysia may well be at the hill slopes of inland areas. Long Akah, by virtue of its location, receives a mean annual rainfall of more than 5000 mm.
3. The northwest coastal areas of Sabah experience a rainfall regime of which two maxima and two minima can be distinctly identified. The primary maximum occurs in October, and the secondary maximum occurs in June. The primary minimum occurs in February, and the secondary minimum occurs in August. Although the difference in the rainfall amounts received during the two months corresponding to the two maxima is small, the amount received during the month of the primary minimum is substantially less than that received during the month of the secondary minimum. In some areas, the difference is as much as four times.
4. In the central parts of Sabah where the land is hilly and sheltered by mountain ranges, the rainfall received is relatively lower than other regions and is evenly distributed. However, two maxima and two minima can be noticed, though somewhat less distinct. In general, the two minima occur in February and August, whereas the two maxima occur in May and October.
5. Southern Sabah has evenly distributed rainfall. The total annual rainfall received is comparable over central part of Sabah. The period February to April is, however, slightly drier than the rest of the year.

MEAN RELATIVE HUMIDITY

Located near the equator, Malaysia generally has quite uniform high relative humidity ranging between 70% and 90%. In 2012, the mean relative humidity for Peninsular

Malaysia was recorded between 77.4% (Petaling Jaya) and 90.8% (Cameron Highlands). The mean relative humidity in Sabah was between 80.3% (Tawau) and 84.4% (Sandakan), whereas mean relative humidity in Sarawak was between 82.0% (Miri) and 85.5% (Bintulu) (Anon 2013).

EVAPORATION

Among all the factors affecting the rate of evaporation, cloudiness and temperature are the two most important factors in this country. These two factors are, however, interrelated. A cloudy day means less sunshine and, thus, less solar radiation resulting in a lower temperature.

An examination of the evaporation data shows that the cloudy or rainy months are the months with lower evaporation rate, whereas the dry months are the months with higher evaporation rate. It is noted that Senai has an average evaporation rate of 2.6 mm/day in the month of November, the lowest for lowland stations. On the other end of the scale, Kota Kinabalu has the highest average evaporation rate of 6.0 mm/day in the month of April. For highland areas such as Cameron Highlands where the air temperature is substantially lower, the evaporation rate is proportionally lower too. Although lowland areas have an annual average evaporation rate of 4–5 mm per day, Cameron Highlands has a rate of only about 2.5 mm per day.

WIND FLOW

The climate on the Peninsula is directly affected by wind from the mainland; hence, it differs from the East, which is more of maritime weather. Malaysia faces two monsoon wind seasons: The southwest monsoon blows from late May to September, whereas the northeast monsoon occurs from November to March. The northeast monsoon brings in more rainfall compared to the southwest monsoon, originating in China and the North Pacific (Anon 2008). The southwest monsoon originates from the deserts of Australia. Months between March and October form transitions between the two monsoons (Marshall Cavendish Corporation 2008).

The wind flow patterns of Malaysia show some uniform periodic changes, although the wind over the country is generally light and variable. Based on these changes, four seasons can be distinguished, namely the southwest monsoon, northeast monsoon, and two shorter periods of intermonsoon seasons (MMD 2014).

The two wind patterns of Malaysia consist of the southwesterly winds, which is generally light below 15 knots and the steady easterly or northeasterly winds of 10–20 knots. The former prevailing wind system forms the southwest monsoon season, which is usually established in the later half of May or early June and ends in September. The latter prevailing wind flow, on the other hand, forms the northeast monsoon season, which usually commences in early November and ends in March. The winds over the east coast states of Peninsular Malaysia may reach 30 knots or more during periods of strong surges of cold air from the north (cold surges).

During the two inter monsoon seasons, the winds are generally light and variable. Between the months of April and November, when typhoons frequently develop over the West Pacific and move westward across the Philippines, southwesterly winds over the northwest coast of Sabah and Sarawak region may strengthen to reach

20 knots or more. Typhoons can sometimes hit Malaysia especially from July to mid November and can cause heavy damage, flooding, and erosion.

As Malaysia is mainly a maritime country, the effect of land and sea breezes on the general wind flow pattern is very marked especially during the days with clear skies. On bright sunny afternoons, sea breezes of 10–15 knots very often develop and reach up to several tens of kilometers inland. On clear nights, the reverse process takes place, and land breezes of weaker strength can also develop over the coastal areas (MMD 2014).

SOIL CHARACTERISTICS

In Malaysia, the environment is generally moderate and does not experience an extreme high or low temperature. The pH values of most of the soils in Malaysia are ranging from 4.5 to 5.5. Accordingly, the soil moisture content in Malaysia is generally at 60%–70% for the whole year. It is therefore necessary to analyze weather, topography, vegetation, types, and other physiological properties of soil with respect to the Malaysian setting.

Fire is another major factor on soil formation. Fire can impact all the vegetative land and can affect the soil directly or indirectly. These effects may be of short term such as decrease in biotic activity, darkening of ground, increased pH, and the formation of water repellent layer affecting the soil moisture greatly. The long-term effects include change in biota, climate change, forming new combustive materials such as charcoal, and reshaping the morphology of soil via erosion (Giacomo and Certini 2001).

In addition to fire, topography or landscape position also affect the moisture content of soil through slope height as shown in Figure 1.2. Steep slopes facing the sun are warmer than the bottom land. It also affects the type and amount of plant growth.

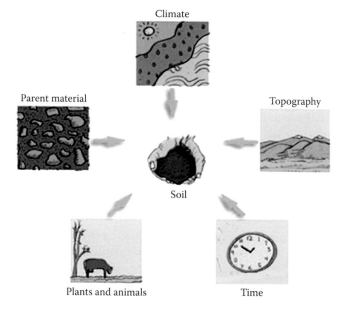

FIGURE 1.2 The soil forming factors.

The Malaysian agriculture is classified by two distinct sectors: the plantation and the smallholders. The major crops cultivated here are oil palm, rice, rubber, coconut, and orchard. In Malaysia, some acid sulfate soil is used for the cultivation of cocoa and oil palm. However, soil acidity and aluminum toxicity are considered to be the main reason for the failure of cocoa crops (Muhrizal et al. 2003). In Malaysia, separate networks of storm water and sanitary sewer are used. Bong et al. (2014) examined the sediment size characteristics and deposition analysis on 24 sediment samples from urban concrete drains in Kuching City.

Nordiana et al. (2014) studied the topographic impact at Bukit Bunuh, Lenggong, Perak, Malaysia by using 2D resistivity method. According to the results obtained by them Bukit Bunuh is highly fractured granitic bedrock with estimated depth of 5–60 m from ground surface. The ground 2D resistivity is related to many geological parameters such as mineral, fluid content, porosity, and degree of water saturation in rock. Tan et al. (2014) elucidate the land use and climate variability in the Johor River basin, Malaysia. They found that land-use change and variation in climate cause the increase in surface water, soil water content, and evaporation capacity. Tropical peatlands cover 27.1 and 2.6 million hectares in Southeast Asia and in Malaysia respectively.

The decomposition of peat soil leads to the emission of CO_2 release into the atmosphere via microbial and root respiration and decomposition of plant residues (Kuzyakov 2006). A group of Choo et al. (2014) partitioned the emission of CO_2 release and dissolved organic carbon from drained pineapple-cultivated peatland at Saratok, Malaysia. They examined partitioning of CO_2 emission with three treatments: (1) bare peat soil, (2) peat soil cultivated with pineapple, and (3) bare peat soil fumigated with chloroform. They found that CO_2 emissions were higher under bare peat (218.8 tonne CO_2 ha/year) than under bare peat treated with chloroform (205 tonne CO_2 ha/year), and they were the lowest (179.6 tonne CO_2 ha/year) under pineapple-cultivated peat soil. This decrease in CO_2 emission under pineapple cultivation was due to the regulation of photosynthesis and soil autotrophic activities. They conclude that CO_2 emission was neither affected by soil temperature nor by soil moisture, but the emission seemed to be controlled by moderate soil temperature fluctuation in the wet and dry seasons.

Colica et al. (2014) studied the effects of microbial secreted exopolysaccharides on hydrological behavior of soil crusts in desert sandy soils. The capability of the crusts to retain moisture, coupled with the preservation of their structural integrity, was found to be positively correlated with the amount of total carbohydrates present in biologically induced soil crusts. The study concluded that microbial secreted carbohydrates increase the water availability in the first layers of sandy soils and reduce water infiltration by protecting the soil from erosion.

ECOSYSTEM

An ecosystem can be defined as a community made up of all living organisms with the environment of their surrounding area that interacts with each other. Ecosystems on earth can be divided into three major classes:

1. Terrestrial ecosystems
2. Marine ecosystems
3. Freshwater ecosystems

TERRESTRIAL ECOSYSTEMS/DRYLAND FOREST

The type of terrestrial ecosystem found in Malaysia is known as tropical rainforest. The tropical rainforests of Western Malaysia start at the borders of Thailand in the north and stretch over the Kra Peninsula, to the straits of Johor in the South (Shuttleworth 1981).

MARINE ECOSYSTEMS

Malaysia's warm tropical sea areas are home to some of the richest coral reefs, mangrove forests, green sea turtles, and other endangered marine species such as hawksbill turtles, dugongs, whale sharks, and humphead wrasse. This vast sea area is rich with fishery resources and habitats (http://www.wwf.org.my/about_wwf/what_we_do/, accessed August 14, 2015).

FRESHWATER ECOSYSTEMS

Freshwater ecosystems are a subset of Earth's aquatic ecosystems. They include lakes and ponds, rivers, streams, springs, and wetlands. They can be contrasted with marine ecosystems, which have a larger salt content (https://en.wikipedia.org/wiki/Freshwater_ecosystem, accessed August 18, 2015). Examples of freshwater ecosystems in Malaysia are Tasik Bera, Tasik Chini, and Tasik Kenyir.

FLORA AND FAUNA OF MALAYSIA

Flora is defined as the plant life occurring in a particular region or time, generally the naturally occurring or indigenous—native plant life. *Fauna* is all of the animal life of any particular region or time. Flora, fauna and other forms of life such as fungi are collectively referred to as biota. Sometimes bacteria and fungi are also referred to as flora (John et al. 2002; Clifford et al. 2014; *Merriam Webster Online Dictionary* 2014). Zoologists and paleontologists use fauna to refer to a typical collection of animals found in a specific time or place, for example, *the butterfly fauna of Sungai Imbak Forest Reserve* (Jalil et al. 2008) or *the fauna and flora of a semideciduous forest in Perlis, Peninsular Malaysia* (Edwards et al. 1996). Paleontologists sometimes refer to a sequence of faunal stages, which is a series of rocks all containing similar fossils.

The flora of Malaysia is estimated to contain around 12,500 species of flowering plants and more than 1,100 species of ferns and fern allies (CBD first report). There are still abundance of flora remains to be known, especially of lower plants such as the bryophytes, algae, lichens, and fungi.

The tropical circumstances and the extensive rainforests of Malaysia have led to a huge diversity of plant and animal species. Surveys identified that there are about 12,500 species of flowering plants, which are inclusive of 2,000 tree species, more than 1,100 species of ferns and fern allies, 800 types of orchids, and 200 types of palms. Malaysia rainforest also houses a well-known plant that is found only in Southeast Asia, the Rafflesia. It occurs mainly in the rainforests of Borneo and Sumatra.

The diversity of fauna in Malaysia is also great. Malaysia is estimated to contain 20% of the world's animal species, and it includes some of the most biodiverse areas on the planet (Alexander 2006). In the vertebrates, there are 300 species of wild mammals, 700–750 species of birds, 350 species of reptiles, 165 species of amphibians, and more than 300 species of freshwater fish. Besides that, it was estimated that there are 1,200 species of butterflies and 12,000 species of moths (CBD first report).

Rain forest fauna includes elephants, orangutans, panthers, seladang (Malaysian bison), Sumatran rhinos, Malaysian sun bear, deer (mouse deer, sambar deer, barking deer), wild pigs, Bornean bearded pigs, tree shrews, honey bears, forest cats, civets, monkeys, crocodiles, lizards, and snakes (Alexander 2006).

Peninsular Malaysia houses three big cats: the Indochinese tiger, the leopard, and also the clouded leopard, although East Malaysia lacks the tiger. The seladang is the largest wild ox in the world and weighs about a ton. An immense variety of insects, particularly butterflies, and some 508 breeding species of birds are also found in Malaysian rainforest.

The flora and fauna of Malaysia can best be described according to the biome, a distinct region occupies by a large community of plants and animals.

TERRESTRIAL FLORA AND FAUNA: TROPICAL RAIN FOREST OF MALAYSIA

About 70% of Malaysia consists of tropical rainforest (Saw 2007), which is believed to be 130 million years old (Richmond 2010). It is composed of a variety of types, although dipterocarps occupy majority of the land area. In Peninsular Malaysia, it is estimated to harbor 8500 species of vascular plants where camphor, ebony, sandalwood, teak, and many varieties of palm trees are found. The forests of East Malaysia consisting 15,000 species of vascular plants are estimated to be the habitat of around 2,000 tree species. They are one of the most diverse areas in the world, with 240 different species of trees occupying every hectare of land area (Anon 2008).

The lowland forests contain some 400 species of tall dipterocarps (hardwoods) and semihardwoods. Dipterocarps can grow to about 50 m tall (Alexander 2006). More than half of the country's surface exists of jungle. The national park Taman Negara contains one of the world's oldest primary rainforest.

Plants easily found in tropical rainforest in Malaysia include heavy hardwoods such as *Intsia bijuga* (merbau), *Shorea albida* (balau), and *Neobalanucarpus heimii* (chengal) and light hardwoods such as kapur, kempas, and *Dipterocarpus acutangulus* (keruing). *Koompassia excelsa* presented in Figure 1.3 is the tallest tree in the rain forest and may grow to a height of 250 ft. Another interesting tree is *Dyera costulata* (jelutong), which supplies sap for chewing gum making. Another group of timber is dangerous and poisonous trees such as *Antiaris toxicaria* (ipoh tree), *Strychnos* spp. (ipoh akar), ipoh kenik, and rengik (Shuttleworth 1981).

1. *Dipterocarpus acutangulus*: *Dipterocarpus acutangulus* presented in Figure 1.4 is a species of tree in the family Dipterocarpaceae. It is found in peninsular Malaysia and Borneo, where it is locally known as *keruing merkah* or *keruing beludu*. It is an emergent tree up to 60 m tall. The tree occurs in mixed dipterocarp forests found on sandy and sandy clay soils on coastal hills and inland ridges, up to 1000 m altitude. It occurs in at least one protected area (Kabili-Sepilok Forest Reserve).

FIGURE 1.3 View of *Koompassia excelsa* (tualang) from the base. (From http://www. rainforestjournal.com/the-tualang-tree-or-koompassia-excelsa/, retrieved March 18, 2015.)

FIGURE 1.4 *Dipterocarpus* spp. found in Tawau Hills Park, Sabah. (From http://sabah-go-green. blogspot.com/2012/09/tallest-tropical-tree-in-world-tawau.html, retrieved March 18, 2015.)

    2. *Rafflesia arnoldii* (corpse flower): This flora is easily found in heath forests (Anon 2008), which host many numbers of *Rafflesia* genus, the largest flower in the world with a maximum diameter of 1 m (The Malaysia Rainforest 2011, http://rainforests.mongabay.com/deforestation /2000/Malaysia.htm.).

        Figure 1.5 shows *Rafflesia*, the plant with the largest flower in the world and can reach a weight of 10 kg. This plant is the symbol of the province

FIGURE 1.5 Blooming Rafflesia found in the forest of Southeast Asia.

Sabah on the island Borneo. However, *Rafflesia* has a strong odor of decaying flesh (http://www.wonderfulmalaysia.com/malaysia-flora-and-fauna. htm).

They also contain a large number of carnivorous plants such as pitcher plants, bladderworts, sundews, and ant-house plants. Carnivorous plants are fascinating in that they capture and digest various preys in order to accumulate nutrients for growth. The habitats where they normally occur are nutrient poor. There are many species of carnivorous plants, and they are found throughout the world. Of the seven genera of pitcher plants, the *Nepenthes* genus is the largest found from Northern Australia throughout Southeast Asia to Southern China. There are also some species found in India, Madagascar, and a few other islands.

3. *Nepenthes* spp.: It is also known as *periuk kera* (monkey pot); these pitcher plants are almost unique to South East Asia and are widely distributed with about six species (Shuttleworth 1981). The pitchers can grow large enough to contain up to 1 L of fluid and are normally found on the ground contrary to most species that have stems that grow up into trees and have pitchers growing off them. This Pitcher plant is famous for its ability to catch and digest rats, frogs, and lizards. Most other *Nepenthes* sp. captures a wide variety of small invertebrates and insects but nothing as large as a rat.

   As seen in Figure 1.6, the habitat where *N. rajah* is found is usually moist, well drained, and has open grassy clearings that are prone to landslides. The soil conditions are always ultramafic. This is soil that is very low in silica content and very rich in iron and magnesium. These photos were all taken in the Mesilau forest preserve at the higher end of their latitudinal habitat at approximately 2600 m.

4. Malayan tiger (*Panthera tigris jacksoni/Panthera tigris malayensis*): Revered as Malaysia's national animal and featured in the country's coat of arms; the Malayan tiger shown in Figure 1.7 is highly native to the jungles of Peninsular Malaysia. It is locally known as *harimau belang*, which literally means *striped tiger* in the Malay language, and is often recognized as a symbol of bravery.

FIGURE 1.6 *Nepenthes rajah.* (From Clark, C., *Nepenthes of Borneo*, Natural History Publications [Borneo] Sdn. Bhd, Malaysia, 1997.)

FIGURE 1.7 Malayan tiger. (From http://hubpages.com/hub/Ten-unique-species-you-must-not-miss-in-Malaysia, retrieved April 17, 2015.)

Due to rapid deforestation and rampant hunting for tiger body parts, which are believed to possess medicinal value, it is now facing the threat of extinction.

5. Asian elephant (*Elephas maximus*): It is found in the tropical rainforests of both Peninsular and Borneo Malaysia; the Asian elephant is another species that is fast becoming extinct due to deforestation and widespread poaching for its tusks. Asian elephant shown in Figure 1.8 is generally smaller than the African elephant, and some of its body parts differ from the latter. In Malaysian Indian culture, just like in India, the Asian elephant is regarded with reverence in Hinduism, being a common depiction of the Hindu deity Lord Ganesha.

6. Sumatran rhinoceros (*Dicerorhinus sumatrensis*): The Sumatran rhinoceros is not only the smallest existing rhinoceros species on earth, it is also

FIGURE 1.8   Asian elephants.

the most endangered species. Sumatran rhinoceros shown in Figure 1.9 was once among the most populous species in the jungles of Peninsular Malaysia and Sabah, but now it is one of the rarest, largely due to excessive poaching for its horn and logging activities that have destroyed much of its natural habitats. It is believed that the only local concentration of the Sumatran rhino population that still exists to this day is found in the Taman Negara or National Park of Malaysia.

FIGURE 1.9   Sumatran rhinoceros. (From http://hubpages.com/hub/Ten-unique-species-you-must-not-miss-in-Malaysia, retrieved April 17, 2015.)

FIGURE 1.10 Malayan tapir. (From http://hubpages.com/hub/Ten-unique-species-you-must-not-miss-in-Malaysia, retrieved April 17, 2015.)

7. Malayan tapir (*Tapirus indicus*): Among the world's four existing species of tapir, the Malayan tapir is the largest by size and the only one native to Asia. It is known locally as *cipan* or *tenuk* in the Malay language and is found primarily in the lowland tropical jungles of Peninsular Malaysia. This animal is unique in the sense that it has a huge white or light-colored patch extending from its shoulders to its rear end, whereas the rest of its body is black or dark colored as shown in Figure 1.10. Its solitary nature and long-gestation period, coupled with extensive deforestation, has rendered the Malayan tapir an endangered species under government protection.

8. Bornean orangutan (*Pongo pygmaeus*): Figure 1.11 shows the Bornean orangutan, which is highly native to the tropical rainforests of Borneo in

FIGURE 1.11 Bornean orangutan. (From http://hubpages.com/hub/Ten-unique-species-you-must-not-miss-in-Malaysia, retrieved April 17, 2015.)

FIGURE 1.12 Proboscis monkey. (From http://hubpages.com/hub/Ten-unique-species-you-must-not-miss-in-Malaysia, retrieved April 17, 2015.)

which it can be found in significant numbers in the dense and mountainous interiors of the East Malaysian states of Sabah and Sarawak.

9. Proboscis monkey (*Nasalis larvatus*): Although not as famous as the orang-utan, the proboscis monkey is another species of primates found only on the island of Borneo (Figure 1.12). In the Malay language, it goes by the name *bekantan*. What make this primate so unique are its unmistakably large nose and bulging stomach that resemble pot bellies. Most of the extant populations of proboscis monkeys today dwell mainly in the rainforests of Kalimantan, Indonesia, but some can still be found particularly in national parks and protected areas throughout Sabah and Sarawak.

10. Rhinoceros hornbill (*Buceros rhinoceros*): Perhaps, it is the most recognizable member of the animal kingdom in the East Malaysian state of Sarawak; this hornbill is definitely no stranger to the people of the state. Being one of the largest extant species of hornbills, the rhinoceros hornbill holds deep significance to the people of Sarawak, being a divine symbol revered in the traditional beliefs of the state's various indigenous tribes. Today, the rhinoceros hornbill shown in Figure 1.13 is recognized as the state bird of Sarawak, in which the state's Malay nickname, *Bumi Kenyalang*, literally

FIGURE 1.13 Rhinoceros hornbill. (From http://hubpages.com/hub/Ten-unique-species-you-must-not-miss-in-Malaysia, retrieved April 17, 2015.)

translates into *land of the hornbills*. Despite being found in larger numbers in Sarawak's rainforests, this attractive bird can also be spotted in several rainforest areas throughout Peninsular Malaysia.

In addition to the rhinoceros hornbill, there exist numerous other species of hornbills native to both Peninsular and East Malaysia's wildlife-rich tropical jungles that you should not miss. These include the great hornbill (*Buceros bicornis*), the oriental pied hornbill (*Anthracoceros albirostris*), the black hornbill (*Anthracoceros malayanus*), and the Malabar pied hornbill (*Anthracoceros coronatus*).

COASTAL FLORA AND FAUNA

Apart from rainforests, Malaysian coastal land displays a large area of mangroves of over 1425 km^2 (Saw 2007) and a large amount of peat forest. Peat bogs are resulted from the sediment buildup along the coastal land that is fringed by the Mangroves. The peat forests of coastal land of the Peninsular Malaysia provide an important habitat for water birds and fish.

Besides, fig trees are also found, attracting small mammals and birds. As altitude increases, oaks, chestnuts, and herbaceous plants, namely, buttercups, violets, and valerian become more numerous, until moss-covered evergreen forests are reached from 1520 to 1830 m (Anon 2008). Malaysia's marine ecosystem includes diverse coral systems (Anon 2014b; Naiker 2014).

Mangrove Forest

Mangrove forests are unique ecosystem that was found along the estuaries and coastal mudflats. There are many species of mangroves, but those mostly present in Malaysia belong to the genera *Rhizophora*, *Avicennia*, *Sonneratia*, *Bruguiera*, and *Carapa*. They are very fast growing plants and able to reproduce themselves in 5 years (Shuttleworth 1981).

In Kuala Selangor National Park, the plants have adopted to cope with the anaerobic and highly saline conditions found in this type of environment. Within the mangrove forest, you will find 4 different families of mangrove trees and a total of 13 species of mangrove trees such as *Bruguiera cylindrica* locally known as Bakau putih (Figure 1.14).

FIGURE 1.14 Bakau putih. (From http://www.natureloveyou.sg/Bruguiera% 20cylindrica/ Main.html, retrieved April 18, 2015.)

FIGURE 1.15   *Boleophthalmus boddarti.* (From https://commons.wikimedia.org/wiki/File:
Boleophthalmus_boddarti_Langkawi_Malaysia%281%29.JPG, retrieved April 18, 2015.)

Varieties of palm tree also grow in the coastal areas of North Borneo such as Nipah, coconut, areca, oil palm, nibong, sago, and traveller's palm. The broad leaves of Nipah palm (*Nipah fruticans*) were widely used in making rooftop (Shuttleworth 1981).

It also provides a protective breeding ground for many varieties of commercially important marine invertebrates including crabs, prawns, and bivalves, which in turn support a large variety of birds and mammals. Species found here include the fiddler crab *Uca rosea*, mudskipper species such as *Boleophthalmus boddarti* (Figure 1.15) and the banded archerfish *Toxotes jaculatrix*.

More than 90 species of migratory and resident birds, 20 fish species, 5 reptile species, and 14 crustacean species have been recorded in Kota Kinabalu Wetland. The mangrove swamp offers habitat for migratory birds such as common kingfisher (*Alcedo atthis*) and common sandpiper (*Actitis hypoleucos*). Almost all year round the most common waterbirds are herons and egrets (Lee 2011).

Marine Fishes, Mammals, and Reptiles

Generally, it was estimated that there were 1500 species of fishes in Malaysian waters (Mazlan et al. 2005). Study conducted by Rumeaida et al. (2014) revealed that a total of 4747 individual fish was observed inside reef areas of Bidong Island from all three stations with 1428 fish of 38 species from Pantai Pasir Cina, 1714 fish of 25 species from Pantai Limau Purut and 1605 fish of 26 species from Teluk Belanga.

It was reported that there are a total number of 29 species of marine mammals in Malaysian waters and at least 13 species of whales, dolphins, porpoises, and dugongs. A further 16 species of cetaceans such as *Orcinus orca*, *Physeter catodon*, and *Kogia breviceps* can also be spotted in Malaysia coastal areas. Common cetacean seen in the region is *Balaenoptera edeni* (Bryde's whale) (Mazlan et al. 2005).

Malaysian coast are also renowned as the habitat for sea turtles. Marine turtles in and around the Malaysian Peninsula include the green sea turtle, leatherback, hawksbill, and olive ridley sea turtle. The population of nesting sea turtles of the Malay Peninsula declined from about 10,000 in the 1950s to fewer than a hundred in 2000. As of 2012, on average, only 10 nests were found at the annual counts (http://animals.mom.me/reptiles-amphibians-malaysian-peninsula-2826.html, accessed August 18, 2015).

FIGURE 1.16 Leatherback turtle (*Dermochelys coriacea*). Length: 127–213 cm (50–84 in.). (From http://global.britannica.com/animal/leatherback-turtle, retrieved April 18, 2015.)

Leatherback turtle: The leatherback turtle (*Dermochelys coriacea*), shown in Figure 1.16, is the largest species of sea turtle and has been attracting tourists to Rantau Abang, a beach on the east coast state of Terengganu, for the last few decades. Every March, when the nesting season begins, thousands of tourists gather at the beach at night to see these giants come ashore and lay their eggs. In the 1960s, there were records of up to 10,000 turtles coming ashore annually (Ministry of Natural Resources and Environment 2006).

Coral Reefs

Coral reefs in Malaysia are estimated to cover close to 4006 km². Coral reefs support not less than 700 species of fish that are dependent on coral reefs as a habitat. Coral reefs are valuable economic and ecological resources. They have important ecosystem functions that provide crucial goods and services to hundreds of millions of people, mostly in developing countries. They are the foundation of a significant proportion of the global tourism industry and are a major source of biodiversity (Ministry of Natural Resources and Environment 2014).

FRESHWATER FLORA AND FAUNA

Freshwater Fishes

Freshwater fishes of Malaysia are diverse and inhabit a great variety of habitats ranging from small torrential streams to estuarine, highly acidic ecosystems and also alkaline waters. Several species are endemic. There are 290 species of freshwater fish in Peninsular Malaysia.

Although inventory figures for Sabah are 100 and for Sarawak are 200, these are believed to be underestimates because their inventories started later compared to Peninsular Malaysia. In Sabah and Sarawak, the focus has mainly been on major rivers. Hence species in isolated, inaccessible, and other inland water bodies have yet to be explored (Ministry of Natural Resources and Environment 2009). Table 1.2 presents the number of freshwater fish species and their production values.

TABLE 1.2
Number of Freshwater Fish in Malaysia

Freshwater Fish Species, 1990s

Total number of species	449	X
Number of threatened species	14	X

Freshwater Seafood Production
Freshwater fish catch {b}

1990 (metric tons)	12,995	3,783,743
2000 (metric tons)	22,636	5,959,055
Freshwater aquaculture production		
1987 (metric tons)	2,738	5,029,515
1997 (metric tons)	20,303	15,469,848

Source: Water Resource Institute (2009). Malaysia—Water Resources and Freshwater Ecosystems Factsheet (WRI), http://www.eoearth.org/view/article/154400, August 17, 2015, Retrieved September 16, 2014.

Tasik Bera

Tasik Bera with 6870 ha of wetlands in the Ramsar site consists of freshwater and peat swamp forest (5440 ha, 79%), open transition swamp forest (510 ha), filled with *Pandanus* plants (Figure 1.17) and *Lepironia articulata* (Figure 1.18) (800 ha, 12%), and open water (120 ha, 2%). There is a very wide range of habitats community consisting of algae and macrophytes.

A total of 374 plant species have been recorded of which 10 species are known to be endemic to Peninsular Malaysia, whereas 328 species of algae have been recorded during the research in the 1970s. Diversity of vertebrate fauna in Tasik Bera wetlands and surrounding forest is in line with the flora diversification. A total of 453 vertebrate species have been recorded consisting of 62 species of amphibians and reptiles, 94 species of fish, 230 species of birds, and 67 species of mammals. Fish species diversity is also a key value for Tasik Bera; offering breeding, treatment, and source of food for fish from the Pahang River (Rahim et al. 2013).

Pandanus odoratissimus (screw pine) also known as *mengkuang* among the locals; the leaves are commonly used as material for making handicrafts. The plant has long leaves with long parallel veins. Along the edge of the leaves are found hard spines, which can prick hands while handling them.

Lepironia articulate shown in Figure 1.18 is a grass-like perennial plant growing from 40 to 200 cm tall. The leaves are harvested from the wild and used for making baskets as a packing material.

FIGURE 1.17 Screw pine. (From http://purplemelastoma.blogspot.com/2010_11_08_ archive.html.)

FIGURE 1.18 *Lepironia articulate.* (From http://www.chm.frim.gov.my/Gallery.aspx? page=172, retrieved April 18, 2015.)

SUMMARY

Malaysia is a tropical country with diverse climate, hydrology, and ecosystem. Being in the equatorial region, Malaysia is very rich both in its flora and fauna, which existence have been well supported by the climate.

REFERENCES

Ahmad, Z. (2010). The Land (West Malaysia). In: *Encyclopaedia Britannica*. Retrieved October 30, 2014, from http://global.britannica.com/place/Malaysia.

Alexander, J. (2006). *Malaysia Brunei & Singapore*. Chatswood: New Holland Publishers. pp. 46–50.

Anon. (2008). Malaysia Travel Guide: Climate of Malaysia. Circle of Asia. Retrieved July 28, 2008, from http://www.circleofasia.com/GeographyandClimateMalaysia.htm.

Anon. (2013). Compendium of Environment Statistics, Malaysia 2013. Department of Statistics, Malaysia. p. 3.

Anon. (2014a). AR4 SYR Synthesis Report Annexes. Ipcc.ch. Retrieved November 15, 2014, from http://www.ipcc.ch/publications_and_data/ar4/syr/en/annexes.html.

Anon. (2014b). Wonderful Malaysia Top 100 Tips eBook. Retrieved October 30, 2014, from http://www.wonderfulmalaysia.com/malaysia-flora-and-fauna.htm.

Bong, C.H.J., Lau, T.L., and Ghani, A.A. (2014). Sediment size and deposition characteristics in Malaysian urban concrete drain a case study of Kuching city. *Urban Water Journal*, 11, 74–89.

Central Intelligent Agency (CIA). (2006). The World Fact Book: Malaysia. Retrieved October 30, 2014, from https://www.cia.gov/library/publications/theworldfactbook/geos/my.html.

Certini, G., (2001). Nitrogen in humic and fulvic acid fractions from a volcanic soil under pine: recovery efficiency of two extractants. *Journal of Plant Nutrition and Soil Science*, 164(3): 267–269.

Certini, G. (2014). Fire as a soil-forming factor. *Ambio*, 43, 191–195.

Choo, L.N.L.K. and Ahmed, O.H. (2014). Partitioning carbon dioxide emission and assessing dissolved organic carbon leaching of a drained peatland cultivated with pineapple at Saratok, Malaysia. *The Scientific World Journal*, 2014, 1–9.

Clarke, C. (1997). *Nepenthes of Borneo*. Malaysia: Natural History Publications (Borneo) Sdn. Bhd.

Clifford, E.S., Villella, R., Morrison, P., and Mathais, J. Sampling the bacterial flora of freshwater mussels. Retrieved August 2, 2014, from http://pubs.usgs.gov/bitn/97-007/pdf/bitn97-007.pdf.

Colica, G., Li, H., Rossi, F., Li, D., Liu, Y., and Philippic, R.D. (2014). Microbial secreted exopolysaccharides affect the hydrological behavior of induced biological soil crusts in desert sandy soils. *Soil Biology & Biochemistry*, 68, 62–70.

Ecologyasia.com. (2010). Mount Kinabalu—Revered abode of the dead. Retrieved October 30, 2014, from http://www.ecologyasia.com/htmlloc/mountkinabalu.htm.

Edwards, D.S., Booth, W.E., and Choy, S.C. (1996). The fauna and flora of a semi-deciduous forest in Perlis, Peninsular Malaysia. In: *Tropical Rainforest Research —Current Issues*. Dordrecht, the Netherlands: Springer. pp. 153–161.

Encyclopædia Britannica. (2010). Main range (mountains, Malaysia). Retrieved October 30, 2014, from http://www.britannica.com/EBchecked/topic/358619/MainRange.

Jalil, M.F., Mahsol, H.F., Wahid, N. and Ahmad, A.H. (2008). A preliminary survey on the butterfly fauna of Sungai Imbak Forest Reserve, a remote area at the centre of Sabah, Malaysia. *Journal of Tropical Biology and Conservation*, 4(1), 115–120.

John, D.M., Whitton, B.A., and Brook, A.J. (2002). *The Freshwater Algal Flora of the British Isles: An Identification Guide to Freshwater and Terrestrial Algae.* New York: Cambridge University Press.

Kiew, R., Chung, R.C.K., Saw, L.G., and Soepadmo, E. (2012). Flora of Peninsular Malaysia. Series II: Seed plants. *Malayan Forest Records*, 49(3), 385.

Kiew, R., Chung, R.C.K., Saw, L.G., Soepadmo, E., and Boyce, P.C. (2010). Flora of Peninsular Malaysia–Series II: Seed plants. *Malayan Forest Records*, 1(49), 329.

Kiew, R., Chung, R.C.K., Saw, L.G., Soepadmo, E., and Boyce, P.C. (2011). Flora of Peninsular Malaysia Series II: Seed plants. *Malayan Forest Records*, 2(49): 329.

Kuzyakov, Y. (2006). Sources of CO_2 efflux from soil and review of partitioning methods. *Soil Biology & Biochemistry*, 38, 425–448

Lee, K.H. (2011). Mangrove fauna and their adaptation in the Kota Kinabalu Wetlands. *Sabah Society Journal*, 28, 15–22.

Malaysian Meteorological Department. 2008. Weather phenomena. (http://web.archive.org/web/20080320105546/http://www.kjc.gov.my/english/education/weather/monsoon01.html). Malaysian Meteorological Department. Archived from the original on March 20, 2008. Retrieved October 30, 2014, from http://www.kjc.gov.my/english/education/weather/monsoon01.html.

Malaysian Meteorology Department, MMD (n.d.). In Ministry of Science, Technology and Innovation (MOSTI). Retrieved November 11, 2014, from http://www.met.gov.my/index.php?option=com_content&task=view&id=75&Itemid=1089.

Marshall Cavendish Corporation. (2008). World and Its Peoples: Malaysia, Philippines, Singapore, and Brunei, New York: Marshall Cavendish Corporation. pp. 1156, 1158–1161, 1164, 1166–1171.

Mazlan, A.G., Zaidi, C.C., Wan Lotfi, W.M., and Othman, B.H.R. (2005). On the current status of coastal marine biodiversity in Malaysia. *Indian Journal of Marine Sciences*, 34(1): 77–86.

Merriam Webster Online Dictionary. (2014). Retrieved September 23, 2014, from http://webster.com/cgi-bin/dictionary?va=flora.

Ministry of Natural Resources and Environment. (2006). *Biodiversity in Malaysia.*

Ministry of Natural Resources and Environment. (2009). Fourth National Report to the Convention on Biological Diversity.

Ministry of Natural Resources and Environment. (2014). Fifth National Report to the Convention on Biological Diversity.

Ministry of Science, Technology and the Environment. (1998). *Malaysia First National Report to the Conference of the Parties of the Convention on Biological Diversity.*

Muhrizal, S., Shamshuddin, J., Husni, M.H.A., and Fauziah, I. (2003). Alleviation of aluminum toxicity in an acid sulfate soil in Malaysia using organic materials. *Communications in Soil Science and Plant Analysis*, 34, 2993–3012.

Naiker, S.S. (2014). Need to conserve natural resources. News Straits Times dated July 11, 2000. Page 5. NASA. 2008. Coral reef image. Retrieved October 30, 2014, from http://www.nasa.gov/vision/earth/lookingatearth/coralreef_image.html.

New Straits Times. (2010). Go: A diver's paradise. Retrieved October 30, 2014, from http://www.nst.com.my/nst/articles/Go_Adiver__8217_sparadise/Article, December 1, 2010.

Nordiana, M.M., Saad, R., Saidin, M., Ismail, N.A., and Jinmin, M. (2014). The resistivity topography in delineating crater rim of meteorite impact at Bukit Bunuh, Lenggong, Perak (Malaysia). *Electronic Journal of Geotechnical Engineering*, 19, 377–388.

Oon, H. (2008). Globetrotter Wildlife Guide Malaysia. London: New Holland Publishers. p. 11.

Parris, B.S., Kiew, R., Chung, R.C.K., Saw, L.G., and Soepadmo, E. (2010). Flora of Peninsular Malaysia–Series I: Ferns and Lycophytes. *Malayan Forest Records*, 1(48), 249.

Rahim, H., Serin, T., and Abdul Wahab, M.A.M. (2013). Tasek Bera Forest Reserve in Pahang: Deplete or conserve? *Economic and Technology Management Review*, 7, 61–70.

Rahman, N.N., Furuta, T., Takane, K., and Mohd, M.A. (1999). Antimalarial activity of extracts of Malaysian medicinal plants. *Journal of Ethnopharmacology*, 64(3), 249–254.

Richmond, S. (2007). *Malaysia, Singapore and Brunei*. Melbourne, Australia: Lonely Planet. pp. 63–64.

Richmond, S. (2010). *Malaysia, Singapore & Brunei*. http://books.google.com/?id=VMKOu zRxOJsC&printsec=frontcover#v=onepage&q&f=false.

Richmond, S. (2010). *Malaysia, Singapore and Brunei*. Melbourne, Australia: Lonely Planet. pp. 74–75, 78–82.

Rumeaida, M.P., Shariff, M.M.D., and Bsdri, F.M.I. (2014). Fish diversity and abundance in Bidong Island, South China Sea, Malaysia. *International Journal of Bioflux Society*, 3(7), 176–183.

Saw, L.G. and Chung, R.C.K. (2007) *Tree Flora of Sabah and Sarawak*. (free online from the publisher, lesser resolution scan PDF versions). Kuala Lumpur, Malaysia: Forest Research Institute Malaysia, http://www.itto.int/files/user/pdf/publications/PD186%2091/pd186-91-4rev1(F)%20e.pdf. pp. 92–93. Retrieved November 22, 2007.

Shaaban, A.J. (1996). *Proceeding: Seminar Hidrologi Dataran Tinggi*, Universiti Teknologi Malaysia (UTM), Johor. 30–31.

Shuttleworth, C. (1981). *Malaysia's Green and Timeless World*. Kuala Lumpur, Malaysia: Heinemann Educational Books (Asia).

Stevens, A.M. (2004). *Kamus Lengkap Indonesia Inggris*. http://books.google.com/?id=cF97 FsuNAC&printsec=frontcover#v=onepage&q&f=false. Athens: Ohio University Press. p. 89.

Tan, M.L., Ibrahim, A.L., Yusop, Z., Duan, Z., and Ling, L. (2014). Impacts of land-use and climate variability on hydrological components in the Johor River basin, Malaysia. *Hydrological Sciences Journal*, 1–28.

The Malaysian Rainforest. 2011, http://rainforests.mongabay.com/deforestation/2000/Malaysia.htm.

Water Resource Institute. (2009). Malaysia—Water Resources and Freshwater Ecosystems Factsheet (WRI). Retrieved September 16, 2014, from http://www.eoearth.org/view/article/154400 on August 17, 2015.

2 Geology and Hydrology

Habibah Jamil, Rizwana Naureen,
and Muhammad Aqeel Ashraf

CONTENTS

INTRODUCTION

Hydrology is generally a scientific study on water. Water plays enormous roles in the physical environment. Studies on water cover various disciplines that concern physical environment. Water exists in various forms and is found in numerous locations of environment. All these complex processes are linked to one another forming what is known as hydrology cycle or water cycle (Wan Ruslan Ismail, DBP 1994). International Scientific Hydrology Association has acknowledged four main disciplines in hydrology:

1. Surface water
2. Groundwater
3. Snow and ice
4. Limnology—pond study

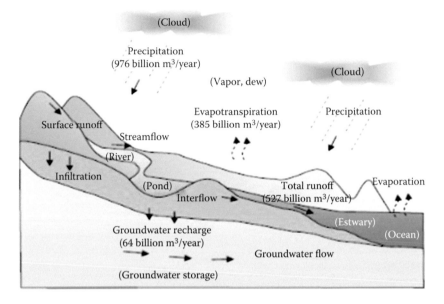

FIGURE 2.1   The hydrological cycle showing average values for Malaysia. (From NWRS 2000.)

Two most studied disciplines are surface water and groundwater. Everything that linked to the phases of the hydrology cycle on the soil surface is closely related to climate. Thus, it is unavoidable to discuss weather and climate whenever we are discussing hydrology. As an example, rainfall is the most important component in hydrology science as a hydrology cycle would not function without it. In other words, anything and everything that are influential to the climate changes will also affect hydrology.

Hydrology cycle refers to the continuous water movement whether on the soil surface or underground. Figure 2.1 explains the quantification of hydrology cycle in Malaysia.

In reality, there is no clear starting or ending point of the cycle. Despite that, the cycle has been broken down into three main processes for better understanding. Those three main processes are as follows:

1. Evaporation
2. Precipitation
3. Process on soil surface

PLATEAU IN MALAYSIA

Out of 330,000 km² of Malaysia's total area, 42% are plateau (*Proceeding of Seminar on Hydrology of Plateau Areas*, July 30–31, 1996). Forestry Department, Malaysia Meteorological Service and Tenaga Nasional Berhad (TNB) or National Electric Utility Co., (Mohamad Sabtu, 1994) (*Proceeding of Seminar on Hydrology of Plateau Areas*, July 30–31, 1996) reported that the classification of an area as plateau is based on the following characteristics:

1. A significant decrease in air temperature occurs at levels exceeding 300 m (above sea level).
2. The onset of the difference of level land with forest hill dipterocarp forest was observed on an area that is 300 m above the sea level.
3. Significant differences in geomorphology occur at the level of 300 m (above sea level).

Lower soil temperature in the plateau area makes it a suitable agricultural land, especially for vegetables and temperate plants.

Malaysia gets an average annual rainfall of around 3000 mm equivalent to 990 billion m^3. As much as 360 billion m^3 (36%) returned to atmosphere through evaporation process, 566 billion m^3 (57%) became runoff, and 64 billion m^3 (7%) was infiltrated into the soil. Out of 990 billion m^3 yearly rainfall, about 50% fall onto the plateau area. Heavy rainfall, steep topography, and low evapotranspiration led to a high rate of water surface runoff and eventually transformed plateau into an area, which is sensitive to land-use change and human activity.

Geographic History

Malaysia is divided into two distinct parts: Peninsular Malaysia at the southern tip of continental Asia and the states of Sarawak and Sabah at the northern part of Borneo Island. Due to different tectonic settings and geological evolution, both Peninsular Malaysia and Sarawak and Sabah depict their own geological characteristics. For example, rocks of Peninsular Malaysia are much older (from Cambrian) compared to that of rocks in Sarawak (from pre-Cretaceous) and Sabah (from pre-Cretaceous). Significant publication on geology of Peninsular Malaysia is that of Gobbett and Hutchison (1973) and Hutchison and Tan (2009), whereas the geology of Sarawak and Sabah has been published in Hutchison (2005). Detail geological evolution of Southeast Asian region, including Peninsular Malaysia, Sarawak, and Sabah, has been published in Hutchison (2007). The geological timescale is shown in Table 2.1.

Peninsular Malaysia has been formed by the collision of the Sibumasu block at the west with the eastern Cathaysian block upon the closure of Paleo-Tethys Sea during Triassic age. The resulted Bentong–Raub suture marked the collision boundary, extending from north to south of Peninsular Malaysia. Following this event, widespread plutonisms formed granitic ranges at the western and eastern belts, which led to the uplift of Peninsular Malaysia. Shallow marine sedimentation occurred at the northwestern and central part of the newly formed landmass. Throughout Jurassic to Cretaceous period, siliciclastic sediments were deposited in diverse terrestrial settings, ranging from alluvial fans, braided rivers, flood plains, and deltas (Abdullah 2009; Burton 1973). Peninsular Malaysia has almost been entirely emergent throughout Cenozoic, and sporadic pull-apart basins have been developed onshore during Tertiary. Sea-level fluctuations throughout Quaternary resulted in the coastal plain deposits that fringed the Peninsular Malaysia.

TABLE 2.1
Geologic Timescale

Era	Period		Epoch	Age (million years ago)
Cenozoic	Quaternary		Holocene	0.01
			Pleistocene	2.6
	Tertiary	Neogene	Pliocene	5.3
			Miocene	23.0
		Paleogene	Oligocene	33.9
			Eocene	56.0
			Paleocene	66.0
Mesozoic	Cretaceous		Late	100
			Early	145
	Jurassic		Late	164
			Middle	174
			Early	201
	Triassic		Late	237
			Middle	247
			Early	252
Paleozoic	Permian			299
	Carboniferous	Pennsylvanian		323
		Mississippian		359
	Devonian			419
	Silurian			444
	Ordovician			485
	Cambrian			540

Source: The Geological Society of America, 2012. The Geologic Time Scale 2012. Boston: Elsevier.

Geological evolution of the northern Borneo, including Sarawak and Sabah, took place from pre-Cretaceous to early Cenozoic of age (Madon 1999; Wannier et al. 2011). The land has been formed by the accretion of oceanic sediments (Hutchison 2007) due to subduction of the Miri microcontinent beneath Borneo. During pre-Cretaceous period, southern Sarawak was part of the Paleozoic Kalimantan continental block, whereas the whole northern region (including Sabah and north Sarawak) was once a deep marine basin, becoming shallower southward. During late Cretaceous to Eocene period, the flysch bed sequences of Rajang Group (Belaga, Mulu, and Kelalan Formations) were deposited in the deep sea basin. At the same time, the deep marine sediments of Sapulut, Trusmadi, and East Crocker Formations were deposited in the troughs, whereas the shallow marine limestone of Madai-Baturong was formed on the highs. Ophiolithic rocks and the associated pelagic sediments (Chert–Spilite Formation) were deposited further east, in the active volcanic deep marine environment.

During Paleocene period, the Luconia block (the drifted Asian continental fragment) has been subducted southwestward beneath the continental core of Borneo, leading to the closure of Rajang Sea and the accretion of oceanic sediments toward the core. Further southwestward collision during the late Eocene *soft collision* resulted to the uplifted, deformed, and piled up of the deep marine sediments, forming a mountainous giant sedimentary prism in Sarawak (Rajang thrust fault) and (Crocker thrust fault). By this time, Sarawak and the central and eastern part of Sabah were uplifted and underwent erosion, supplying sediments to the shallow marine and intermontane basins including Baram Delta complex. Although most of the region was uplifted into shallow marine, the deep marine environment persisted at the west coast of Sabah.

During Oligocene to Early Miocene period, the spreading of South China Sea marginal basin pushed Miri microcontinent southwestward to Borneo. The Late Miocene *hard collision* resulted to the deformation and uplift of the younger deep marine sediments in Sabah, flanking the older sedimentary prism. The ophiolithic rocks have been obducted and imbricated with the associated deep marine sediments (Chert–Spilite Formation), and the uplifted rocks were eroded, supplying various lithologies and sizes of clasts for olistostrome deposits. During that time, olistostrome deposition was accompanied by explosive volcanisms. The under-thrusting was also responsible for the intrusion of Mount Kinabalu (Late Miocene) and other smaller granitic stocks around Mount Kinabalu area. Continuous uplift of the Crocker Range and the east coast area resulted to the shallow marine depositional basin in central Sabah. In Sarawak, the Miri Zone was uplifted into shallow marine and estuarine environments, allowing the deposition of shallow marine siliciclastic sediments and calcareous rocks. The progressively southward-driven Miri microcontinent rendered the anticlockwise rotation of the north Borneo to the present-day landform.

The opening of the Celebes Sea and the movement of Palawan plate (southwestward) and Sulawesi plate (northwestward) led to the formation of rift depression on the older rock basement, allowing the deposition of fluvio-deltaic sediments at the east and southeastern areas of Sabah. During Pliocene epoch, rocks in Sarawak underwent compressional deformation that resulted in a series of folds and reverse faults. By this time, northern Sarawak and Sabah have been uplifted to the present-day landforms. Active faulting continues in Sabah, allowing the deposition of marine sediments together with volcanisms at eastern Sabah.

GEOLOGY

PENINSULAR MALAYSIA

The geology of Peninsular Malaysia can be divided into three belts: the western, central, and eastern belts according to its distinct stratigraphy as presented in Figure 2.2. Bentong–Raub suture separates the western belt (Sibumasu block) from the central and eastern belts (Cathaysian block). The western belt is mainly occupied by the metasedimentary rocks of the shallow and deep marine environments. Rocks of the northwestern Peninsular Malaysia constitutes the Paleozoic shallow marine (Machinchang Jerai, Setul, and Chuping Formations) and deep marine (Mahang, Singa, and Kubang Pasu Formations) sediments, and the Mesozoic shallow marine

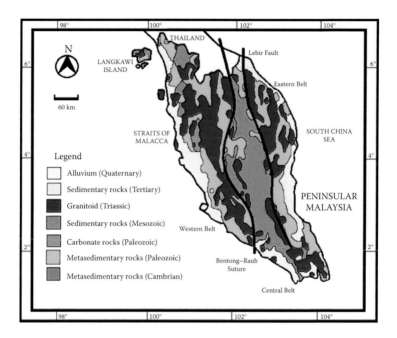

FIGURE 2.2 The geological map of Peninsular Malaysia. (From Geological Survey of Malaysia 1988.)

(Kodiang Formation) and deep marine (Semanggol Formation) sediments. To the south, this belt is occupied by the Paleozoic metasedimentary rocks (phyllite and schist), and the carbonate rocks in Kinta Valley and Kuala Lumpur areas (Gobbett and Hutchison 1973; Hutchison and Tan 2009).

The eastern belt lies to the east of Lebir Fault and is mainly overlain by the Paleozoic shallow marine sediments (argillites and arenites) with some isolated limestone lenses, volcanic rocks, and conglomerates. The Mesozoic thick succession of continental strata (Gagau Group) occupies the northeast coast of Peninsular Malaysia. The central belt lies between the Bentong–Raub suture (western boundary) and the Lebir Fault (eastern boundary). This belt is predominantly occupied by the Paleozoic argillaceous and volcanic rocks, with smaller occurrence of arenaceous and calcareous sediments, with intermittent submarine volcanisms. Sporadic limestones occur as linear belts on both sides of the belt. The Mesozoic rocks in this belt are rich in pyroclastic rocks. Gua Musang Formation is a predominant calcareous and argillaceous rock with subordinate arenite, pyroclastic, and lava flow, whereas Semantan Formation is a predominant carbonaceous shale interbedded with rhyolithic tuff (Gobbett and Hutchison 1973; Hutchison and Tan 2009).

Peninsular Malaysia has been intruded by a series of parallel igneous plutonic rocks of Triassic age, with minor occurrence of Cretaceous intrusions. The main range biotite granite at the west of the suture is distinctly megacrystic and of the ilmenite series, whereas the granitoids of the eastern belt are more equigranular, containing both biotite and hornblend, and mainly of the magnetite series

(Hutchison 2007). The sporadic occurrences of onshore Tertiary basins are depicted in Peninsular Malaysia (Raj et al. 2009a).

Peninsular Malaysia is fringed with lowland coastal plains ranging from a few kilometers to 30 km in width. Soils of the coastal area constitute sandy soils, marine clay, peat, and associated riverine alluvial soils (Isahak 1992). Sandy soils are mainly found along the east coast of the peninsula due to the strong currents operating in the South China Sea. On the other hand, the calmer Straits of Malacca brings in fine sediments to the west coast that resulted in the predominant marine clay deposits.

SARAWAK

Geologically, Sarawak can be divided into three main zones: the Kuching, Sibu and Miri Zones representing different lithologies and ages (Hutchison 2007; Madon 1999; Wannier et al. 2011). Each zone is separated by tectonic lines known as the Lupar Line (Kuching–Sibu Zones boundary) and the Tatau–Mersing Line (Sibu–Miri Zones boundary). Lupar Line is characterized by the Lubok Antu Melange (Early Eocene), a complex mixture of boulders (sandstone, radiolarian chert, basic igneous rocks, limestone, and serpentinite) in strongly sheared chloritized pelitic matrix. The melange has been formed in a trench basin upon the closure of Rajang Sea during Late Cretaceous age. The Tatau–Mersing Line constitutes a structurally complex zone of ophiolithic rocks (spilite, basalt, tuff, and radiolarian chert). The geology of Sarawak is depicted in Figure 2.3.

Sarawak is mainly covered by the metamorphosed sedimentary rocks. Generally, rock sequences in Sarawak are stratigraphically younger and less deformed toward

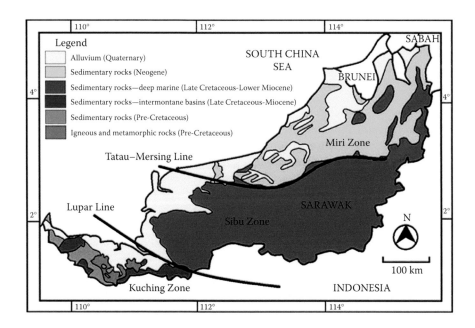

FIGURE 2.3 Geological map of Sarawak. (Modified from Geological Survey of Malaysia 1992.)

the northeast. The oldest rock is Kerait Schist (Upper Carboniferous) at Kuching Zone, whereas the youngest is the unconsolidated sediments of Liang Formation (Pliocene to Middle Pleistocene) at Miri Zone.

Kuching Zone lies from the southwestern Sarawak to the Lupar Line suture zone. This zone is covered by the highly deformed metamorphic rocks of Kerait Schist and Tuang Formation (Hutchison 2007), before they are unconformably overlain by the shallow marine Terbat Formation (Late Carboniferous to Early Permian). All of these rocks are overlain by shallow marine to estuarine and swamp deposits of Sadong Formation (Middle Triassic to Late Triassic) and the contemporaneous pyroclastic rocks of Serian Volcanic (Late Triassic). Overlying these rocks are the thick siliciclastic sequences and minor carbonate rocks of Jurassic to Cretaceous age. These sequences and the older rocks are intruded by the Middle Jurassic to Late Cretaceous plutonic suites, before they are overlain by the shallow marine to terrestrial siliciclastic rocks (Kayan and Silantek Formations, and Plateau Sandstone) of Late Cretaceous to Middle Miocene age (Madon 1999).

Sibu Zone is characterized by the crescenic belt of low-grade metamorphosed and intensely folded rocks of Rajang Group (Late Cretaceous to Eocene), with minor igneous rock intrusions and pillow lavas. Rajang Group constitutes the highly deformed flysh-like deep-sea sediment, known as the Lupar and Belaga Formations.

Miri Zone constitutes the northward extension of Rajang Group, locally known as Kelalan Formation (Late Paleocene to Early Eocene). These rocks are overlain unconformably by the Oligocene to Miocene shallow marine to deltaic sedimentary rocks (Tatau, Nyalau, Setap and Younger Formations), which in turn are overlain unconformably by the shallow marine to estuarine sediments of Lambir, Miri, and Tukau Formations. Overlying Miri Formation is the coastal to terrestrial deposits of Seria (Late Miocene) and Liang (Pliocene to Middle Pleistocene) Formations. Alluvial Quaternary sediments occupy a vast area of Lupar Valley and northern part of Sarawak.

Sabah

The geology of Sabah can be divided into two main blocks: the Sabah Basin and the Sabah Complex Block as given in Figure 2.4. Sabah Basin lies at the west coast and northern Sabah, which are overlain predominantly by the Paleogene to Neogene imbricated deep marine sedimentary rocks, represented by the mountainous Crocker and Trusmadi Ranges. Sabah Basin constitutes predominantly the folded, thrust, and uplifted turbidite flysch deposits, locally known as the Sapulut, Trusmadi, and East Crocker Formations. These formations are the eastward extension of the uplifted flysh-like deposit at Sibu Zone (Belaga Formation), which extend along the eastern margin of Miri Zone continental margin (Kelalan Formation), to reach the Mount Kinabalu area.

The East Crocker, Sapulut, and Trusmadi Formations are flanked by the younger deep-sea sedimentary rocks (sandstone and mudstone) known as the West Crocker, North Crocker, Main Crocker, and Temburong Formations (Late Paleogene to Early Neogene). Argillaceous strata of the North Crocker Formation predominantly occupy the northern Sabah, whereas the arenaceous strata of the West Crocker.

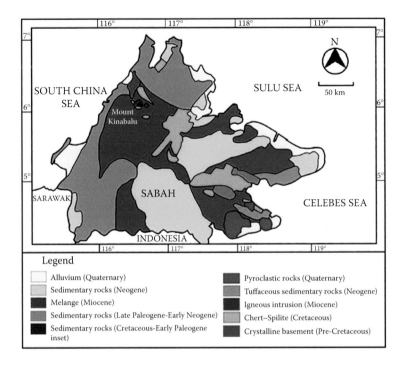

FIGURE 2.4 Geological map of Sabah. (Modified from Leong, K.M., *The Petroleum Geology and Resources of Malaysia.* Petroliam Nasional Berhad (PETRONAS), Kuala Lumpur, 1999; Hutchison, C.S., *Geological Evolution of South-East Asia.* Geological Society of Malaysia, Kuala Lumpur, 2007.)

Formation occupy the western and northwestern Sabah. Further in the south, rocks of the West Crocker Formation become more argillaceous with thin siltstone layers and locally known as the Temburong Formation (Oligocene to Lower Miocene). The Main Crocker Formation (Eocene to Early Miocene) constitutes a thick sequence of rhythmically interbedded coarse and fine clastics of complex tectonic structure, representing submarine fan to the eastern margin of the Miri Zone continental margin (Leong 1999; Hutchison 2007).

These rocks are unconformably overlain by the younger shallow marine sedimentary rocks. To the north in Kudat Peninsula, the rhythmic sandstone and shale alternations pass into massive grain flow and sandstone fan deposits known as the Kudat Formation (Oligocene to Early Miocene). Shallow marine sediments occupy the central (Labang and Kuamut Formations) and eastern Sabah (Tanjung, Kalabakan, and Kapilit Formations). Overlying these rocks are the shallow marine to deltaic and estuarine sediments of the Pliocene to Pleistocene age (Simenggaris, Liang and Timohing Formations).

The plutonic intrusion of Mount Kinabalu (1401 m in height) forms the highest peak in Borneo Island. The central part of this mountain constitutes hornblend adamellite, whereas the outer shell is of adamelite porphyry. Mount Kinabalu (Middle to Late Miocene), together with other smaller plutonic stocks, is distributed at the east of Rajang Thrust Fault and have intruded the older Trusmadi Formation and the ophiolitic ultrabasic rocks underneath (Hutchison 2007).

Sabah complex block occupies the eastern and southeastern areas of Sabah, constitutes the crystalline basement, Sabah complex or Chert–Spilite Formation, the volcanic and pyroclastic rocks, melange (olistostrome), and shallow marine deposits.

The pre-Cretaceous core of crystalline basement forms the oldest rock in Sabah. The metamorphic and igneous complex of this basement is distributed in the eastern Sabah. Sabah complex or Chert–Spilite Formation (Late Cretaceous to Late Eocene) constitutes a complex sedimentary (radiolarian chert, limestones, and clastic sediments) and volcanic rocks (extrusive ultrabasic and basic igneous rocks). Chert–Spilite Formation is mainly exposed at the eastern and central Sabah, with smaller occurrences at Mount Kinabalu and Kudat areas. These rocks present as an arcuate belt extending from eastern Sabah to the central and northwestern area of Kudat Peninsula.

In eastern Sabah, the uplifted ophiolite is overlain by the extensive Early Miocene melange and olistostrome deposits. These deposits (Early Miocene) constitute various sizes (pebbles to huge boulders) of assorted older rock lithologies (chert, limestone, gabbro, spilite, sandstone, and serpentinite) in marine clay matrix. Melange deposits are locally known as Kuamut, Garinono, Ayer, and Kalabakan Formations in eastern Sabah, whereas in western Sabah, it is known as Wariu Formation.

The pyroclastic and andesitic volcanic rocks (Late Miocene) in eastern Sabah are known as Tungku Formation. Shallow marine deposits are represented by the Late Oligocene to Late Miocene Kalumpang Formation (predominant mudstone and shale, with minor occurrence of siliciclastic sediments and tuff) and Gomantong Limestone. The Quaternary alluvium deposits in Sabah constitute volcanic ash and tuff, mudstone, shale, thinly bedded sandstone, and coal of paralic to shallow marine depositional environments.

HYDROLOGY

Malaysia is located in the equatorial zone (1°N–7°N and 100°E–119°E) and experiences the humid tropic climate, characterized by uniform temperature, high humidity, and copious rainfall. In 2012, the temperature of this country ranges from 13.0°C to 36.9°C. The mean relative humidity for Peninsular Malaysia, Sarawak, and Sabah are 77.4%–90.8%, 82.0%–85.5%, and 80.3%–84.4%, respectively. Malaysia naturally has abundant sunshine and thus has solar radiation. On the average, Malaysia receives about 6 hours of sunshine per day. The annual evaporation rate at the lowland areas are 4–5 mm per day, and the annual rainfall is 1833.8–3936.2 mm (Department of Statistics Malaysia 2013; Government of Malaysia, Department of Irrigation and Drainage 2009a; Malaysian Meteorological Department 2013).

The amount of rainfall received in both Peninsular Malaysia and the states of Sarawak and Sabah is mainly influenced by seasonal variation known as the southwest and northeast monsoons. The southwest monsoon prevails in the latter half of May or early June and ends in September. During this season, prevailing wind flow

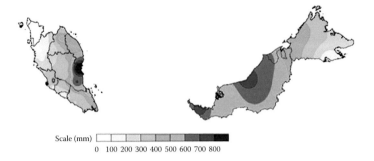

Scale (mm)
0   100 200 300 400 500 600 700 800

FIGURE 2.5 Rainfall map in Malaysia during the northeast monsoon season in December 2013. (From Malaysian Meteorological Department, Official Website Malaysian Meteorological Department [MetMalaysia], Ministry of Science, Technology and Innovation [MOSTI], 2013, http://www.met.gov.my/index.php?option=com_content&task=view &id=75& Itemid=1089&limit=1&limitstart=2, accessed December 1, 2014.)

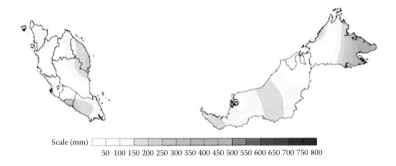

Scale (mm)
50   100 150 200 250 300 350 400 450 500 550 600 650 700 750 800

FIGURE 2.6 Rainfall map in Malaysia during the southwest monsoon season in July 2014. (From Malaysian Meteorological Department, Official Website Malaysian Meteorological Department [MetMalaysia], Ministry of Science, Technology and Innovation [MOSTI], 2013, http://www.met.gov.my/index.php?option=com_content&task=view&id=75&Itemid=1089& limit=1&limitstart=2, accessed December 1, 2014.)

is generally southwesterly and light, below 15 knots. During the northeast monsoon in the month of November to March, the prevailing northeasterly wind is stronger (10–20 knots) due to strong surges of cold air from the north (cold surges). The rainfall maps in Malaysia during the northeast and southwest monsoon seasons are depicted in Figures 2.5 and 2.6, respectively.

The east coast of Peninsular Malaysia receives maximum rainfall in the months of November, December, and January, whereas June and July are the driest months. The southwestern Peninsular Malaysia receives maximum rainfall in the month of October and November, whereas the minimum rainfall is in February. The rest of the peninsula experiences the intermittent two periods of maximum rainfalls (October–November and April–May) and minimum rainfalls. At the northwestern region, the primary minimum occurs in the month of January–February with the secondary minimum in the month of June–July. Elsewhere, the primary minimum occurs in the month of June and July, whereas the secondary minimum occurs in the month of February.

Coastal areas of Sarawak and northeast Sabah receive the heaviest rainfall during the northeast monsoon in the months of December–March. Rainfall in the inland areas of Sarawak is evenly distributed, with slightly less rainfall during southwesterly monsoon (June to August). At the northwest coastal area of Sabah, the maximum rainfall is in the month of October and June, whereas the minimum rainfall is in the month of February and August. Rainfall in the central and southern areas of Sabah is evenly distributed.

The country is blessed with abundant rainwater, which flows into the rivers, lakes, and eventually to the sea. The subsurface water infiltration flows as underground streams or accumulated as groundwater aquifers. The average annual rainfall in Malaysia is equivalent to 990 billion m^3 of water resources, representing 566 billion m^3 (57%) of surface runoff, 64 billion m^3 (7%) of groundwater recharge, and 360 billion m^3 (36%) are lost by evatranspiration (Alam et al. 2010; Japan International Cooperation Agency 1982; Toriman et al. 2010). The amount of surface runoff in Peninsular Malaysia, Sarawak, and Sabah are 147 billion m^3, 306 billion m^3, and 113 billion m^3, respectively (Toriman and Mokhtar 2012).

DRAINAGE

The drainage system in Peninsular Malaysia has evolved since the land emergence and is likely to have been established at the time of maximum marine transgression in the Middle Miocene (Raj 2009b). The drainage patterns are mainly associated with the rocks' competence and sheared rocks along fault zones. Generally, the drainage pattern in Peninsular Malaysia is more or less in north–south and east–west direction. In granitic region, the drainage is mainly rectangular due to the occurrence of joints and faults, whereas radial drainage is depicted in the isolated granitic peaks (Tjia 1973). Rivers on the west coast of Peninsular Malaysia debouch to the Straits of Malacca, and rivers on the east coast area flow into the South China Sea. In the mountainous areas of Sarawak and Sabah, trellis drainage is depicted at the zigzag ridges and valley patterns, associated with the alternate competent sandstones and shales. Rivers in Sarawak and northern Sabah are flowing into the South China Sea, whereas those on the east coast Sabah are flowing into the Sulu and Celebes Seas.

Waterfalls and rapids are common at the upstream areas as water flows on the highly competent rocks such as granite and quartzite. For example, waterfall at Telaga Tujuh (Seven Wells in Langkawi Island depict the erosional surface on the Gunung Raya granitic rocks. The waterfall is named after the occurrence of various sizes of circular erosional features (potholes), resembling cooking pots on the bedrock as shown in Figure 2.7. As river flows downstream into the lowland areas, hydraulic energy decreases allowing sedimentation and siltation at the lower valley region. Upon entering the vast coastal plain areas, rivers at the west coast of Peninsular Malaysia depict meandering drainage patterns before debouching into the Straits of Malacca (Figure 2.8). Similar features are also depicted in the northwestern coastal area of Sarawak. Oxbow lakes, resulted from the bypassed channel, are also common in these areas.

(a)   (b)

FIGURE 2.7   Waterfall at Telaga Tujuh (a) and a pothole on an erosional surface (b).

FIGURE 2.8   The meandering Langat River passing through a flat coastal plain area before debouching into the Straits of Malacca.

Lakes

Lakes in Malaysia are natural and artificial in origin. Natural lakes are not common in this country due to the lack of internal drainage pattern. All of the rivers in Peninsular Malaysia are flowing downstream from the mountainous and hilly areas into either the Straits of Malacca (at the west coast) or the South China Sea (at the east coast). In Sarawak, most of the rivers are flowing downstream from the southern mountainous area to the South China Sea. In Sabah, rivers from the mountainous areas flow down to the South China, Sulu, or Celebes Seas. Natural lakes mainly occur as a substantial freshwater body in the inland valley, limestone hollow, or freshwater swamps. On the other hand, most of the artificial lakes in this country are created from the dam construction for various purposes, and some of them are resulted from the abandoned tin mines.

Natural Lakes

Tasik Bera (Bera Lake) is the largest natural freshwater swamp in Peninsular Malaysia. Being located in the valley between the main and Eastern mountain ranges, the elongated lake measures approximately 27 km in length and 5 km in width. The lake and its surrounding area are a home for Semelai branch of the indigenous people and have been protected under the Ramsar Convention since November 1994. However, the livelihood in Tasik Chini area has been threatened by the severe land use due to increasing rubber and oil palm plantations since the last few decades. The ecosystem of Tasik Bera has been continuously studied to maintain its sustainability (Othman et al. 2014).

Tasik Chini (Chini Lake) is the second largest natural lake in Peninsular Malaysia and is a home for Jakun branch of the indigenous people. Covering an area of approximately 6150 ha, the lake is one of the UNESCO Biosphere Reserve. The elongated lake is formed by the debris-blocked tributary channels during landslides (Idris et al. 2005; Ooi 1968). Over the past few years, the shallow lake ecosystem has been deteriorated by severe sediment loading (30.57 kg/km²/day) due to logging, forestry, and agricultural activities (Sujaul et al. 2013) and due to other activities such as the construction of small dams for recreational and fishing activities. This lake has been continuously monitored and studied, and measures have been taken to restore its sustainability (Gasim et al. 2006).

Tasik Dayang Bunting (Lake of the Pregnant Maiden) offers a picturesque lake on the Dayang Bunting Island, nearby the Langkawi Island. The lake occupies a deep limestone hollow, measuring about 0.8 km in length and 0.4 km in width and is separated from the sea by a thin limestone wall (Jones 1981). This lake, together with other karst landscapes in the island, is part of the Dayang Bunting Marble Geoforest Park (Leman et al. 2007).

Loagan Bunut (Bunut Lake) is the largest natural lake in Sarawak, covering an area of approximately 650 ha and is inhabited by Berawan Community. The lake is located in the Loagan Bunut National Park (LBNP) at the upper reaches of Bunut River between Tinjar and Teru Rivers. The ephemeral lake is seasonally fed by the Tinjar River stream water. During drought season, when the Tinjar River water level is low, the lake has been completely drained, leaving a huge expanse of dried and cracked mud. Upon the rainy season, the shallow Loagan Bunut is once again refilled with water. Logging activities in the watershed area have tremendously increased the sediment load up to 62.5 times from 1980 to 2006 (Tuen et al. 2006). Loagan Bunut has been monitored by the Conservation and Sustainable Use of Tropical Peat Swamp Forests and Associated Wetland Ecosystems Project, supported by the UNDP/GEF and Malaysian Government to protect and conserve its lake ecosystem (Sayok et al. 2009).

Artificial Lakes

Most of the artificial lakes in Malaysia have been created by damming the catchment areas for various purposes including hydroelectric power generation, water supply, irrigation, and flood mitigation (Government of Malaysia, Department of Irrigation and Drainage 2009b). Freshwater reservoirs have been created in the state of Kedah by the construction of the Pedu (catchment area of 171 km^2) and Ahning (catchment area of 122 km^2) Dams in Kedah River Basin, and by the construction of the Muda (catchment area of 984 km^2) and Beris (catchment area of 116 km^2) Dams in Muda River Basin. In Penang Island, two dams have been constructed to serve domestic and industrial water supplies. For water supply in Pahang state, a dam has recently been built in the Kuantan River Basin.

Potable water supplies are highly demanded in the most populated and highly developed area of Klang Valley, especially in the metropolitan city of Kuala Lumpur. Hence, most of the artificial lakes in the Selangor state are mainly created for this purpose. Two dams, the Klang Gates (catchment area of 77 km^2) and Batu Dam (catchment area of 50 km^2), have been constructed at the upper reaches of Klang River Basin to supply water for the Klang Valley residence. More water reservoirs have been created to serve other residential areas by the construction of the Selangor River (catchment area of 197 km^2) and Tinggi River (40 km^2) Dams in Selangor River Basin, and the Langat (41.1 km^2) and Semenyih (56.7 km^2) Dams in Langat River Basin.

Besides potable water supply, freshwater reservoirs also serve as water resources in the cropland areas. Traditionally, paddy farmers irrigate their cropland by pumping or diverting water from the nearby streams. However, upon the advent of double cropping of rice cultivation during 1960s and 1970s, adequate water resources are needed for the off-season crops. To address this matter, the government evolved a policy to confine irrigation schemes at the designated granary areas in Malaysia. The scheme involves the stream water diversion or regulation from the watershed areas to paddy fields. Major canal systems in major river basins including Perak River (Trans Perak Irrigation Scheme), Kelantan River (KEMUBU Irrigation Scheme), and Besut River (KETARA Irrigation Scheme) have been established to serve this purpose. In Sabah, the Penampang Irrigation Scheme has been established to irrigate the cropland at the west coast area (Toriman and Mokhtar 2012).

Tasik Kenyir (Kenyir Lake) is the largest artificial lake in Peninsular Malaysia and Southeast Asia. Covering an area of approximately 369 km^2, the lake serves as hydroelectric power generation known as the Sultan Mahmud Power Station. The lake has been created in 1985 by the construction of Kenyir Dam (155 m in height) in Terenganu River Basin. Besides hydropower, this lake also serves as an alternate gateway to the Taman Negara (The National Park) and as flood mitigator (Yusoff 2007).

Tasik Temenggor (Temenggor Lake) is the second largest hydroelectric reservoir in Peninsular Malaysia. The lake has been created by the construction of Temenggor Dam (127 m of height) at the upper reaches of Perak River. Tasik Timah-Tasoh (Timah-Tasoh Lake) in the state of Perlis has been built in 1992 to generate electricity. Tasik Batang Ai (Batang Ai Lake) is the only artificial lake in Sarawak and has been constructed for hydroelectric power generation (Sovacool and Bulan 2012).

Abandoned tin mines are common features at the west coast of Peninsular Malaysia, particularly in the state of Perak and Selangor. In 1900s, Peninsular Malaysia was once among the largest tin producers in the world of which most of the tin mines were operated in these two states (Fox 1974). However, tin mining operations have decreased since 1980s due to the falls of tin price market. Most of the mines have been abandoned, leaving hundred hectares of marginal land with patches of isolated lakes. As the time goes by, some of these lakes have been transformed into recreational parks for public use. For example, the Taiping Lake Garden in Perak state was formerly a huge tin mine, before it was transformed into a nature sanctuary in 1880 (Thani et al. 2015).

Tasik Putrajaya (Putrajaya Lake) has been created as a natural cooling system at the new public administration center in the Federal Territory of Putrajaya. The 650 ha lake also serves as recreational park and as the venue for water sports competitions including the F1 Powerboat Championship and the Asian Canoeing Championship.

GROUNDWATER

In Malaysia, the annual groundwater recharge and the groundwater aquifers are roughly estimated as 64 billion m^3 and 5000 billion m^3, respectively. The groundwater utilization is 0.156 billion m^3/year, accounting for 3% of the total water usage in this country. Water consumption is mainly for domestic purposes (60%) and for the industrial (35%) and agricultural (5%) sectors (Japan International Cooperation Agency 1982; Razak and Abd. Karim 2009). By the year 2009, the groundwater exploitation in Peninsular Malaysia, Sarawak, and Sabah is approximately 0.2 million m^3/day, 0.05 million m^3/day, and 0.02 million m^3/day, respectively (Razak and Abd. Karim 2009). Groundwater exploration and exploitation throughout the country have been carried out by the Mineral and Geoscience Department of Malaysia (JMG), private sectors, and local residence.

Groundwater resources in Peninsular Malaysia have been tapped from the alluvium (sand and gravel) and hard rock aquifers (limestone, the fractured sandstone, metamorphic, volcanic, and granitic rocks). Alluvium aquifers at the coastal plain of Peninsular Malaysia are the most productive, yielding from 50 to 100 m^3/hour/well. Groundwater resources from alluvium aquifers have been widely used by the residence of north Kelantan as early as 1935 (Heng and Singh 1989). On the other hand, groundwater consumption in the state of Selangor, of particular the Langat Basin, is mainly for industrial purposes, followed by the domestic, agricultural, and commercial usages (Japan International Cooperation Agency and Mineral and Geoscience Department Malaysia 2002). Although limestone is the most productive hard rock aquifer (yield up to 50 m^3/hour/well), groundwater development is unfavored here due to the unsuitable environmental condition (located in the developed areas and may cause sinkhole). Hard rock aquifers are also present in the fractured, sheared, and porous sedimentary and volcanic rocks (can yield 30 m^3/hour/well) and in the jointed, fractured, and weathered granitic rocks (can yield 20 m^3/hour/well) (Razak and Abd. Karim 2009).

In Sarawak, groundwater has been utilized by the residence since 1954, and remain as the main water supply in several coastal villages. The groundwater resources have been tapped from the shallow (sand, gravel, and peat) and deep aquifers (sedimentary basins and fractured hard rocks). Groundwater utilization in Sabah is limited to the isolated villages for domestic usage of which the groundwater resources are from the recent alluvium aquifers (Razak and Abd. Karim 2009).

GEOMORPHOLOGY

The geomorphology of Malaysia has resulted from the combination of rock-type formations, geological structures, and the climatic condition throughout the landmass emergence since Cenozoic era. Rocks of Peninsular Malaysia and the states of Sarawak and Sabah have experienced humid tropic condition and prolonged subaerial exposure. Such condition favored the formation of deep weathering profile and laterite over different bedrock types (Jusop and Ishak 2013; Khairuddin and Madon 1999; Raj 2009b). Intense chemical weathering has led to relatively rapid soil formation, high denudation rate, and various landscape formation including mountainous ranges and development of denser drainage per unit area than the similar rock types under temperate climatic condition (Raj 2009b; Tjia 1973). Sea-level fluctuations throughout Quaternary age and the occurrences of depositional and erosional processes resulted in the development of coastal plains of the peninsula, river terraces, inland plains, and infilled valley. Waves and tides dominate along the coastal areas giving rise to various coastal features (Raj 2009b; Tjia and Syed Abdullah 1992 and 2013).

Geomorphology of Malaysia can be described under a few categories: the highland areas (mountainous range and rolling to undulating terrain), the lowland areas (alluvial plain), coastal features, and the limestone areas (karst landform).

HIGHLAND AREAS

Mountainous ranges and the rolling to undulating landforms represent the denudational terrain over a variety of bedrock resulted from prolonged weathering and erosional processes. Highland areas of Peninsular Malaysia are marked by the Nakawan, Kedah-Singgora, Bintang, Kledang, Main, Benom, Tahan, and East Coast Ranges (Raj 2009b; Tjia 1973), with Mount Tahan as the highest peak in Peninsular Malaysia (2191 m in height). The north–south elongated ranges are alternating with the lower depositional terrain in the form of inland plains, infilled valleys, flood plains, and river terraces. These ranges are flanked by the low convex hills of metasedimentary rocks or by the deeply weathered granitic rocks. Isolated hills such as Mount Jerai and Mount Ledang are formed by the granitic stock intrusions. Jagged ridges on the top of Mount Machinchang (750 m in height) in Langkawi Island have resulted from the successive beds of varying competence. A series of strike ridges and valleys has formed on the alternate competent sandstones with the relatively less competent mudstones shown in Figure 2.9.

The mountainous ranges of Rajang Thrust Faults dominates the highland areas of Sarawak. The ranges extending from northwest to the southeast before curving upside to the northeastern part of Sarawak. The highest ranges are located at the Malaysia–Indonesia border at the south Sarawak. The mountainous ranges extend further to the east, dividing the Sabah state in half by the Crocker Thrust Fault. Jagged features on top of these ranges have resulted from the alternation of the competent sandstones with less resistant mudstone. Besides these mountainous belt, the northwestern and western Sabah are also covered with the mountainous Crocker Range.

FIGURE 2.9 Background view of jagged mountainous range of Mount Machinchang in Langkawi Island.

The granodiorite intrusion of Mount Kinabalu is the highest peak in Southeast Asia (4010 m in height). During the Pleistocene Last Glacial Maximum (LGM), the mountain peak was covered with glacier, indicated by the glacial abrasion morphologies including striations, grooves, chatter marks, roche moutonnee, cirques, and crescentic gouges as shown in Figure 2.10. Tilloid deposits (unconsolidated boulder, gravel, sand, and mud) of Pleistocene age are also distributed at a lower topography (3355 m above the sea level) to the north and south of this mountain (Tjia 1984).

FIGURE 2.10 Glacial abrasion features on top of Mount Kinabalu, Sabah.

LOWLAND AREAS

The flat coastal areas at the western and eastern coast of Peninsular Malaysia have been formed by sedimentation during the Mid-Holocene higher sea-level stand (Bosch 1988; Tjia and Syed Abdullah 1992 and 2013). Following the sea-level regression, the emergence of marine deposits formed a vast area of alluvial plain, whereas the once isolated limestone islands became mogote hills on the plain (Figure 2.11). The former raised shorelines (up to 6 m above present sea level) are indicated by various morphological features including the beach ridges, coquina beds and sea notches on the higher ground, and the river-incised valleys at further inland. Figure 2.12 shows the sea notches.

Northwestern Sarawak is covered by the extensive coastal plains and peat swamps, and drained by Rajang River, the longest river in Sarawak (536 km in length). A large cuspate delta system of Baram Delta (300 km in width) occupies the northeastern area since the Late Miocene time.

FIGURE 2.11 Isolated limestone hills rise on the alluvial plain at the northwestern Peninsular Malaysia.

FIGURE 2.12 Sea notches at the base of limestone hill.

FIGURE 2.13   Beach rock at the southeastern Peninsular Malaysia.

COASTAL FEATURES

Besides sea-level fluctuations throughout the Quaternary period, coastline of Peninsular Malaysia has been shaped by waves and shoreline current (Teh 1992). The straight shorelines, with parallel well-developed beach ridges at the east coast peninsula, have been attributed to the strong waves of the South China Sea. Wave-dominated fan-shaped delta is common, particularly at the mouth of the Kelantan and Pahang Rivers. The old raised shorelines are indicated by abrasion terraces, clinging dead oysters on granitic bedrocks, and beach rocks as shown in Figure 2.13. On the other hand, shorelines at the west coast peninsula are irregular and drained by large river mouth or estuaries. The beach ridges are smaller and narrower compared to that of the eastern coastal areas due to the weaker wave action in the Straits of Malacca (Tjia 1973).

KARST MORPHOLOGY

Karst landscapes including the steep-sided hills (tower karst or mogote), rugged rock pinnacles, and solution features (karren, doline, uvala, poljes, and swallow holes) are commonly found on carbonate bedrocks. Due to bedrock solution by percolating meteoric water, the limestone hills are commonly honeycombed by various sizes of tunnels, caves, and subterranean passages, whereas the surfaces are often deeply pitted. Inside the cave, irregular features ranging from the enormous stalagmites and stalactites to the thin denticular encrustations are decorating the wall and roof cave (Jones 1981; Tjia 1973). Panoramic view of isolated mogote hills (approximately 600 m in height) that rise on the flat countryside can be seen in carbonate bedrock areas throughout the peninsula (Tjia 1973).

Sarawak has spectacular karst landscapes at Subis Complex (Niah National Park) and Melinau Formation (Gunung Mulu National Park). The Niah Cave complex constitutes several voluminous high-ceilinged chambers, roof openings, sinkholes, and roof collapse blocks litter on the cave floor. The cave entrance is encrusted with

FIGURE 2.14 Stalactitic tuffas hanging on the entrance roof to Niah's Great Cave.

stalactite tuffas and columns (Figure 2.14). The Mulu giant karst system constitutes oversize caves, connected by a long network of galleries and subterranean rivers. Deer Cave in Mulu is the largest natural chamber in the world (170 m in width and 120 m in height), characterized by various resurgences, high roof collapse block, and absence of speleothem. Lang Cave, however, depicts various sorts of speleothems including dripstones, straws or stalactites, stalagmites, columns, calcitic curtains, and stalactitic tuffas (Wannier et al. 2011).

CONCLUSION

Due to specific geological evolution, the type of rocks in Peninsular Malaysia is different and much more older compared to that of the rocks in northern Borneo (the states of Sarawak and Sabah). The present landforms in Peninsular Malaysia and northern Borneo have resulted from the uplifted rocks during the Cenozoic era and Pliocene epoch, respectively, followed by intense chemical weathering and erosional processes. Landscapes of this country have been attributed to the combination of rock-type formations, geological structures, and the climatic condition throughout the aerial landmass exposure. Experiencing the humid tropic condition, Malaysia receives abundant rainfall in which the freshwater occurs as the surface and groundwater streams, or confined in the lakes (natural and artificial) and groundwater aquifers (alluvium and hard rock).

REFERENCES

Abdullah, N. T. (2009). Mesozoic stratigraphy. In Hutchison, C. S. & Tan, D. N. K. (Eds.). *Geology of Peninsular Malaysia*. Kuala Lumpur, Malaysia: The University of Malaya and The Geological Society of Malaysia.

Alam, M. et al. (2010). Socioeconomic profile of farmer in Malaysia: Study on integrated agricultural development area in north-west Selangor. *Agricultural Economics and Rural Development*. 7(2), 249–265.

Bosch, J. H. A. (1988). The Quaternary Deposits in the Coastal Plains of Peninsular Malaysia. Geological Society of Malaysia. Quaternary Geology Section, Report QG/1.1988: 81.

Burton, C. K. (1973). Mesozoic. In Gobbett, D. J. and Hutchison, C. S. (Eds.). *Geology of the Peninsular Malaysia: West Malaysia and Singapore*. New York: Wiley-Interscience.

Department of Statistics Malaysia. (2013). Compendium of Environment Statistics Malaysia 2013. ISSN 1511-3396.

Fox, W. (1974). *Tin: The Working of a Commodity Agreement*. London: Mining Journal Books Limited.

Gasim, M. B. et al. (2006). Hydrology: Water quality and land-use assessment of Tasik Chini's feeder rivers, Pahang, Malaysia. *GEOGRAFIA Online Malaysian Journal of Society and Space*. 2, 72–86.

Geological Survey of Malaysia. (1988). *Geological Map of Peninsular Malaysia*. 8th edition. Kuala Lumpur: Geological Survey of Malaysia.

Geological Survey of Malaysia. (1992). *Geological Map of Sarawak*. 2nd edition.

Gobbett, D. J. and Hutchison, C. S. (1973). *Geology of the Peninsular Malaysia: West Malaysia and Singapore*. New York: Wiley-Interscience.

Government of Malaysia, Department of Irrigation and Drainage. (2009a). DID Manual. Volume 4. Hydrology and Water Resources. Available from: http://forum.mygeoportal. gov.my/smanre/aduan/Volume%204_Hydrology%20and%20Water%20Resources.pdf (Accessed: April 2, 2015).

Government of Malaysia, Department of Irrigation and Drainage. (2009b). DID Manual. Volume 5. Irrigation and Agricultural Drainage. Available from: http://forum. mygeoportal.gov.my/smanre/aduan/Volume%205_Irrigation%20and%20Agricultural %20 Drainage.pdf. (Accessed: April 2, 2015).

Heng, T. E. and Singh, M. (1989). Groundwater supply studies in northern Kelantan. *Bulletin of the Geological Society of Malaysia*. 24, 13–26.

Hutchison, C. S. (2005). *Geology of North West Borneo: Sarawak, Brunei and Sabah*. Amsterdam, the Netherlands: Elsevier B.V.

Hutchison, C. S. (2007). *Geological Evolution of South-East Asia*. 2nd edition. Kuala Lumpur, Malaysia: Geological Society of Malaysia.

Hutchison, C. S. and Tan, D. N. K. (Eds.) (2009). *Geology of Peninsular Malaysia*. Kuala Lumpur, Malaysia: The University of Malaya and The Geological Society of Malaysia.

Idris, M., Hussin, K. and Mohamad, A. L. (Eds.) (2005). *Sumber Asli Tasik Chini*. Bangi, Malaysia: Penerbit Universiti Kebangsaan Malaysia.

Isahak, A. (1992). Coastal soils of Peninsular Malaysia. In Tjia, H. D. and Syed Abdullah, S. M. (Eds.). *The Coastal Zone of Peninsular Malaysia*. Bangi, Malaysia: Penerbit Universiti Kebangsaan Malaysia.

Japan International Cooperation Agency and Government of Malaysia. (1982). National Water Resource Study, Malaysia.

Japan International Cooperation Agency and Mineral and Geoscience Department Malaysia. (2002). The study on the sustainable groundwater resources and environmental management for the Langat Basin in Malaysia. Final Report, Volume 1. CTI Engineering & OYO Corporation.

Jones, C. R. (1981). *The Geology and Mineral Resources of Perlis, North Kedah and the Langkawi Islands*. 17, pp. 242–245. Kuala Lumpur: Geological Survey of Malaysia District Memoir.

Jusop, S. and Ishak, C. F. (2013). *Weathered Tropical Soils: The Ultisols and Oxisols*. Serdang, Malaysia: Universiti Putra Malaysia Press.

Khairuddin, M. and Madon, M. (1999). *Malaysia in brief. In PETRONAS. The Petroleum Geology and Resources of Malaysia*. Kuala Lumpur, Malaysia: Petroliam Nasional Berhad (PETRONAS).

Leman, M. S. et al. (2007). Geopark as an answer to geoheritage conservation in Malaysia: The Langkawi Geopark case study. *Bulletin of the Geological Society of Malaysia.* 53, 95–102.

Leong, K. M. (1999). Geological Setting of Sabah. *The Petroleum Geology and Resources of Malaysia.* Kuala Lumpur, Malaysia: Petroliam Nasional Berhad (PETRONAS).

Madon, M. (1999). Geological Setting of Sarawak. *The Petroleum Geology and Resources of Malaysia.* Kuala Lumpur, Malaysia: Petroliam Nasional Berhad (PETRONAS).

Malaysian Meteorological Department. (2013). Official Website Malaysian Meteorological Department (MetMalaysia). Ministry of Science, Technology and Innovation (MOSTI). Available from: http://www.met.gov.my/index.php?option=com_content&task=view&id=75&Itemid=1089&limit=1&limitstart=2. (Accessed: December 1, 2014).

Ooi, J. B. (1968). *Land, People and Economy in Malaya.* London: Longman.

Othman, Z. et al. (2014). Radiometric dating of sediment records in Tasik Bera. *International Journal of Environment, Society and Space.* 2(1), 1–17.

Raj, J. K. et al. (2009a). Cenozoic Stratigraphy. In Hutchison, C. S. and Tan, D. N. K. (Eds.). *Geology of Peninsular Malaysia.* Kuala Lumpur, Malaysia: The University of Malaya and The Geological Society of Malaysia.

Raj, J. K. (2009b.) Geomorphology. In Hutchison, C. S. and Tan, D. N. K. (Eds.). *Geology of Peninsular Malaysia.* Kuala Lumpur, Malaysia: The University of Malaya and The Geological Society of Malaysia.

Razak, Y. A. and Abd. Karim, M. H. A. (2009). Groundwater in the Malaysian Context. *Groundwater Colloquium 2009. Groundwater Management in Malaysia: Status and challenges.* Kuala Lumpur, Malaysia: Akademi Sains Malaysia.

Review of National Water Resources Study (2000–2050) and Formulation of National Water Resources Policy, Volume 1–14, 2011.

Sayok, A. K., Lau, S. and Belanda, R. D. (2009). Impacts of Land-use on Bunut Lake, Sarawak, Malaysia. 13th World Lake Conference, Wuhan, China, 1–5 November. Available from: http://wldb.ilec.or.jp/data/ilec/WLC13_Papers/S15/s15-4.pdf (Accessed: April 24, 2015.)

Sovacool, B. K. and Bulan, L. C. (2012). Energy security and hydropower development in Malaysia: The drivers and challenges facing the Sarawak Corridor of Renewable Energy (SCORE). *Renewable Energy.* 40, 113–129.

Sujaul, I. M. et al. (2013). Hydrological assessment and suspended sediment loading of the Chini Lake catchment, Pahang, Malaysia. *International Journal of Water Resources and Environmental Engineering.* 5(6), 303–309.

Teh, T. S. (1992). The permatang system of Peninsular Malaysia: An overview. In Tjia, H. D. and Syed Abdullah, S. M. (Eds.). *The Coastal Zone of Peninsular Malaysia.* Bangi, Malaysia: Penerbit Universiti Kebangsaan Malaysia.

Thani, S. K. et al. (2015). Public awareness towards conservation of English landscape at Taiping Lake Garden, Malaysia. *Procedia Social and Behavioral Sciences.* 168, 181–190.

The Geological Society of America. (2012). The Geologic Time Scale 2012. Boston: Elsevier.

Tjia, H. D. (1973). Geomorphology. In Gobbett, D. J. and Hutchison, C. S. (Eds.). *Geology of the Malay Peninsula: West Malaysia and Singapore.* New York: John Wiley & Sons Inc.

Tjia, H. D. (1984). *Aspek Geologi Kuaternari Asia Tenggara.* Bangi, Malaysia: Penerbit Universiti Kebangsaan Malaysia.

Tjia, H. D. and Syed Abdullah, S. M. (Eds.). (1992). *The Coastal Zone of Peninsular Malaysia.* Bangi, Malaysia: Penerbit Universiti Kebangsaan Malaysia.

Tjia, H. D. and Syed Abdullah, S. M. (Eds.). (2013). *Sea Level Changes in Peninsular Malaysia. A Geological Record.* Bangi, Malaysia: Penerbit Universiti Kebangsaan Malaysia.

Toriman, M. E. and Mokhtar, M. (2012). Irrigation: Types, sources and problems in Malaysia. In Teang, S. L. (Ed.). *Irrigation Systems and Practices in Challenging Environments.* InTech. Available from: http://www.intechopen.com/books/irrigation-systems-and-practices-in-challengingenvironments/irrigation-types-sources-and-problems-in-malaysia. (Accessed April 24, 2015).

Toriman, M. E. et al. (2010). Issues of climate change and water resources in Peninsular Malaysia: The case of northern Kedah. *The Arab World Geographer.* 12(1–2), 87–94.

Tuen, A. A. et al. (Eds.) (2006). *Scientific Journey Through Borneo. Loagan Bunut: A Scientific Expedition on the Physical, Chemical, Biological, and Sociological Aspects.* Kepong, Malaysia: Peat Swamp Forest Project, UNDP/GEF Funded (MAL/99/G31).

Wannier, M. et al. (2011). Geological excursions around Miri, Sarawak. *1910–2010: Celebrating the 100th Anniversary of the Discovery of the Miri Oil Field.* Miri, Malaysia: EcoMedia Software.

Yusoff, F. M. (2007). Tasik Kenyir: Thermal Stratification and Its Implications on Sustainable Management of the Reservoir. Paper presented at the Colloquium on Lakes and Reservoir Management: Status and Issues. August 2–3, 2007, Putrajaya, Malaysia.

3 Landuse for Agriculture in Malaysia

Wan Noordin Daud and Shafar Jefri Mokhtar

CONTENTS

INTRODUCTION

Agriculture sector is an important sector in the economy of Malaysia. During 1950s, the agricultural sector contributed about 46% of the gross domestic product (GDP) and 80.3% of the employment. However, increasing industrialization with improving country's economy in 1980s resulted in the reduction of agricultural activity. Although role of agricultural sector is decreased, it is still having superiority in the formulation of government policy. Agricultural sector still have major continued importance in earning foreign exchange through exports of palm oil, rubber, and fruits, through contribution to employment, and through ensuring food security for the nation. The National Agrofood Policy (NAP) 2011–2020 is the latest policy focusing on three main objectives: to ensure adequate food supply and food safety, to develop the agrofood industry into a competitive and sustainable industry, and to increase the income level of target groups.

Agricultural production is proportional to land availability and its fertility. Agricultural land refers to the share of land area that is arable, under permanent crops, and under permanent pastures. Nowadays, with increasing population and urbanization, landuse competition is high. This resulted in decreasing of arable land for agriculture. Agriculture activity has to shift into less arable, infertile and problem areas such as steep slope area and tin-tailing soils. Soils in the tropics is well known as highly weathered soils with low pH, low nutrient content and less capability in retaining nutrients supplied.

MALAYSIAN AGRICULTURE LANDUSE

Located near the equator, Malaysia's climate is categorized as equatorial, being hot and humid throughout the year. The average rainfall is 250 cm (98 in.) a year, and the average temperature is 27°C (80.6°F). These climatic conditions are ideal for the growth and production of the tropical perennial tree crops. In geographic point of view, Malaysia can be considered as mountainous with more than half of the land over 150 m above sea level, thus limiting the area suitable for agriculture activity as shown in Table 3.1.

Malaysia has a land area of 33.03 million ha with about 31% of it being arable. Agriculture is one of the main landuse in Malaysia (Aminuddin et al., 1990) and has been a significant contributor to economic development of the country as a source of food, employment, export earner, and raw materials for agro-based industries (Arshad et al., 2007). Agriculture activities in Malaysia are mainly paddy cultivation, plantation crops, fruits, vegetables, cash crops, herbs, spices, flower, aquaculture activities, and livestock production. General landuse of agricultural activities in Malaysia is shown in Table 3.2.

TABLE 3.1
Landuse in Malaysia

Item	Area (million ha)	Percentage (%)
Total land area	33.03	100
Land area suitable for agriculture	10.31	31.2
Land area unsuitable for agriculture	22.72	68.8

Source: Olaniyi, A.O. et al., *Bulg. J. Agri. Sci.*, 19, 60–69, 2013.

TABLE 3.2
Agriculture Landuse in Malaysia for 2013

Landuse Type	Hectare	Percentage (%)
Paddy	674,332	8.99
Plantation crops	6,316,406	84.25
Minor plantation crops	149,765	2.00
Fruits	197,533	2.63
Vegetables	67,777	0.90
Cash crops	19,270	0.26
Herbs and spices	7,159	0.10
Flower	2,227	0.03
Aquaculture	38,835	0.52
Livestock area	23,872	0.32
Total	7,497,176	100

Source: Ministry of Agriculture, 2014; Department of Agriculture, 2014.

Landuse for agriculture in Malaysia is dominated by plantation crops, occupying 6,316,406 ha or 84.25%.

THE PLANTATION AND SMALLHOLDER SUBSECTORS

Agriculture sector in Malaysia can be divided into plantation subsector and smallholders subsector. Plantation could be defined as a land for agriculture activity that is larger than 40.5 ha (100 acres). The plantation subsector is highly commercialized and managed efficiently by the professionals (UPM, 2010). Plantations in Malaysia are usually owned by public-listed corporate entities (e.g., Sime Darby), public land development agencies (e.g., FELCRA), and private companies (e.g., IOI Plantation). These companies are only involved in the production of plantation crops such as oil palm, rubber, and cocoa. Smallholders subsector is less commercialized, and the average farm size is 1.45 ha. It is estimated that there are 1,033,065 smallholders in the country, and the crops grown include plantation crops, rice, fruits, and vegetables.

PLANTATION CROPS

Malaysian agriculture landuse has been dominated by perennial plantation crops, led by oil palm, followed by rubber, cocoa, pepper, and tobacco as shown in Table 3.3. These crops were known as main plantation crops. Landuse for these crops alone occupies more than 80% of the agricultural land area. There are few crops, known as other plantation crops, namely, coconut, coffee, nipa palm, areca nut, roselle, sago, sugarcane, tea, and mushroom.

Oil palm (*Elaeis guineensis*) occupies the largest area of crops grown in Malaysia and its product, palm oil is the biggest foreign exchange earner among all agricultural commodities and products. Oil palm is originated from Sierra Leone, Africa and is grown as plantation crops in Malaysia in 1917 at Tenamarran Estate, Batang Berjuntai, Selangor (UPM, 2010). Cultivation of oil palm is mostly done by the

TABLE 3.3
Planted Hectarage of Main Plantation Crops in 2013

Item	Planted Area (ha)	Production (tonnes)	Average Yield (tonnes/ha)
Rubber	1,057,271	826,421	0.8
Palm Oil	5,229,739	19,216,459	3.7
Cocoa	13,728	2,809	0.2
Pepper	15,130	26,500	1.8
Tobacco	538	–	–
Total	6,316,406	19,325,880	

Source: Ministry of Agriculture and Agro-Based Industry, Malaysia, Agrofood statistics 2013. MOA, Putrajaya, Malaysia, 2014.

FIGURE 3.1   Oldest rubber tree in Malaysia planted at Kuala Kangsar in 1877.

plantation companies. Establishment of Federal Land Development Authority (FELDA) has made new changes in oil palm plantation sector with an objective to eradicate poverty and to increase economic status of people. Through FELDA, large forest areas have been opened for planting of oil palm by landless settlers. Cultivation cycle for oil palm is about 20–25 years. Oil palm can be harvested after 25–30 months after planting. Currently, large plantation companies dominated the oil palm cultivation, with value of 60%. Balance of 40% oil palm is grown by smallholders on their own or under schemes such as Federal Land Consolidation and Rehabilitation Authority (FELCRA) and Rubber Industry Smallholders Development Authority (RISDA) (UPM, 2010).

The first major plantation crop introduced into Malaysia was rubber (*Hevea brasiliensis*). The home of rubber tree is Brazil. One hundred and forty years ago, there was not even a single rubber tree in Malaysia. Seeds collected from Brazil were shipped to Kew Gardens, England and then to the Singapore Botanic Garden in 1877. The first batch of rubber trees was planted in 1877 at Bukit Residency, Kuala Kangsar as presented Figure 3.1. Now, there are 500 million rubber trees growing in Malaysia (Noordin, 2013). The rubber industry has been a pillar of the Malaysian economy since 1950s and continues to be a major contributor until the present day.

Although planted area under rubber has been continuously declining since 1982, natural rubber production remained at about 1 million tonne since 2004, which is indicative of land productivity as shown in Figure 3.2. The importance of natural rubber in terms of socioeconomics cannot be denied as it sustains the livelihood of more than 200,000 smallholder families, whereas the downstream manufacturing sector provides employment to over 64,000 workers; this sector made a significant contribution to the economy with the total export earnings growth from RM5.3 billion

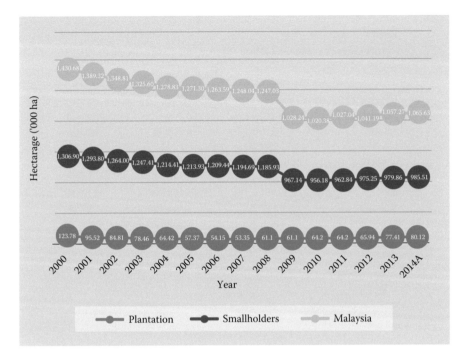

FIGURE 3.2   Rubber planted area in Malaysia from 2000 to 2014.

in 1990 to RM25.3 billion in 2010 (Malaysian Rubber Board, 2015). Due to government's favorable industrial policies, the rubber industry has diversified over the years from planting (upstream) into downstream manufacturing. Today, Malaysia is world's No. 1 exporter of natural rubber gloves, catheters, condom, and latex thread.

Cocoa has been commercially planted since 1950s, whereas cocoa processing began in 1970s. Most of the planting areas are located in Sabah, but most of the processing is based in Peninsular Malaysia. The planting sector has been slowing down over the years, in contrast with processing sector, which has grown tremendously.

Other plantation crops, sometimes known as minor plantation crops, occupy smaller planted area. Coffee is mainly grown by smallholders particularly in the state of Selangor and Johor. Most of coffee grown is *Liberica*, favored by local consumers. Other types of coffee such as *Robusta* and *Arabica* were only about 5%.

Tea plant in Peninsular Malaysia is originated from China. Tea cultivation becomes important in the early twentieth century although it has been brought into Malaysia since the seventeenth century. However, the tea grown commercially is mostly originated from India (Table 3.4). The variety of tea, known as Assam variety, is grown in Cameron Highlands, Pahang (UPM, 2010).

Other than plantation crops, the next important crop in Malaysia is rice as it is the staple food of most Malaysians and is considered a strategic crop. About 501,090 ha of rice is planted in Peninsular Malaysia and 173,242 ha in Sabah and Sarawak as presented in Table 3.5. In the last 3 years, the total area planted decreased slightly, but the average of yield increased steadily as presented in Figure 3.3 and Table 3.6.

TABLE 3.4
Planted Hectarage of Other Plantation Crops in 2011–2013

Type of Crops	Planted Area (ha)		
	2011	2012	2013
Coconut	106,312.40	100,996	87,974.30
Coffee	5,140.50	4,277.40	3,764
Nipa Palm	91.2	29.3	29.2
Areca Nut	341.1	222.3	44.7
Roselle	116.4	86.4	103
Sago	55,901	56,468.90	55,122.90
Sugarcane	4,098.90	4,345.70	237.6
Tea	2,459.40	2,379.70	2,299.30
Mushroom	0	142.90	189.6
Total	174,460.90	168,949	149,764.60

Source: Department of Agriculture, Industrial crops statistics Malaysia 2013, DOA, Putrajaya, Malaysia, 2014c.

TABLE 3.5
Hectarage and Average Yield of Paddy in 2013

State	Planted Area (ha)	Average Yield of Paddy (kg/ha)
Johor	2,960	4,527
Kedah	210,327	4,228
Kelantan	56,280	3,825
Melaka	2,783	3,578
N. Sembilan	1,986	4,242
Pahang	10,357	3,039
Perak	81,636	4,411
Perlis	52,085	4,828
P. Pinang	25,564	5,677
Selangor	37,833	6,280
Terengganu	19,279	4,213
Sabah	38,982	3,050
Sarawak	134,260	1,890
Malaysia	674,332	3,879

Source: Department of Agriculture, Paddy statistics of Malaysia 2013, DOA, Putrajaya, Malaysia; Ministry of Agriculture and Agro-Based Industry, Malaysia. (2014). Agrofood statistics 2013. MOA, Putrajaya, Malaysia, 2014d.

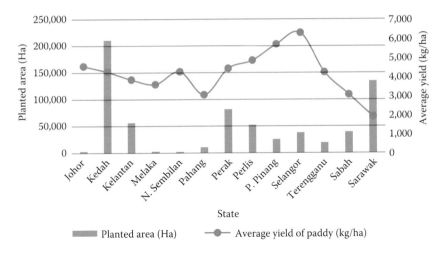

FIGURE 3.3 Hectarage and average yield of paddy in 2013.

TABLE 3.6
Hectarage and Average Yield of Paddy for Year 2011–2013

Year	Planted Area (ha)	Average of Yield (kg/ha)
2011	687,940	3,748
2012	684,545	3,797
2013	674,332	3,879

Source: Department of Agriculture Paddy statistics of Malaysia 2013, DOA, Putrajaya, Malaysia, 2014d.

Most of the area is located in the alluvial coastal regions. Rice has been a traditional crop grown by small farmers.

Currently, rice production is concentrated in eight granary areas in Peninsular Malaysia. In these areas, various inputs such as irrigation, drainage, and rice mills are provided, and rice is grown in two or more seasons in a year. The eight granary areas are as follows:

1. Kuala Muda area in Kedah/Perlis (Muda Agricultural Development Authority [MADA])
2. Alluvial plains of Kemubu area, Kelantan (Kemubu Agricultural Development Authority [KADA])
3. Seberang Perai, Penang
4. Kerian/Sg Manik, Perak
5. South west Selangor
6. Kemasin-Semarak-plains of Kemasin River, Bachok
7. Besut, Terengganu (The ministry of agro-based industries [KETARA])
8. Seberang Perak

The eight main granaries contributed about 70% of the national rice production. Paddy is a highly subsidized crop in Malaysia as the cost of production is relatively higher than in neighboring countries. Subsidies are given in term of seeds, fertilizers, herbicides, and insecticides. Other than that, the prices for the rice produced by farmers are guaranteed and controlled for the consumers.

Table 3.7 shows that fruit crops occupy the largest cultivated area of food crop in the country after paddy with over 197,000 ha. The fruit industry is dominated by smallholders, involving 270,000 farmers. Fruits can be divided into seasonal fruits and nonseasonal fruits. Seasonal fruits include durian, rambutan, mangosteen, pulasan, dokong, duku, and langsat. State of Johore has largest area of fruit crops (42,056.5 ha) with production of 508,812.7 Mt in 2013 as presented in Table 3.8.

TABLE 3.7
Hectarage and Production of Fruits in 2013

Type of Fruit	Planted Area (ha)	Harvested Area (ha)	Production (Mt)
Seasonal fruits	132,896.10	95,385.60	618,551.40
Nonseasonal fruits	64,637.20	53,258.40	894,626.70
Total	197,533.30	148,644.00	1,513,178.10

Source: Department of Agriculture, Fruit crop statistics Malaysia 2013, DOA, Putrajaya, Malaysia, 2014a.

TABLE 3.8
Hectarage and Production of Fruits According to State in 2013

State	Planted Area (ha)	Harvested Area (ha)	Production (Mt)
Johore	42,056.5	35,265.4	508,812.7
Kedah	11,213.2	9,026.5	78,780.8
Kelantan	26,344.1	18,044.2	170,954
Malacca	6,400.1	6,156.3	41,876.4
Negeri Sembilan	8,208.6	7,094.6	80,858.3
Pahang	25,222.5	18,465.6	115,325.6
Perak	13,116.7	10,784	109,805.9
Perlis	966.1	517.9	3,892.5
Penang	5,627.2	3,800.9	27,740.4
Selangor	2,397	2,379	28,471.6
Terengganu	6,162	5,210.6	47,664
Sabah	16,362.5	11,780.7	128,150.4
Sarawak	33,115.9	19,783	169,851.4
W.P. Labuan	341	335.3	994.2
Total	197,533.4	148,644	1,513,178.2

Source: Department of Agriculture, Fruit crop statistics Malaysia 2013, DOA, Putrajaya, Malaysia, 2014a.

According to Table 3.9, as per the type of fruits crops cultivated, banana occupies the largest area compared to other fruits crop. The major fruits being exported by Malaysia are watermelon, papaya, starfruit, and durian. As shown in Table 3.10, in the last 3 years, the area under fruits crop has been reduced from 219,870 ha in 2011 to 197,533 ha in 2013.

TABLE 3.9
Hectarage and Production of Major Fruits in 2013

Type of Fruits	Planted Area (ha)	Harvested Area (ha)	Production (Mt)
Starfruit	845.3	706.6	11,782.70
Papaya	1,934.00	1,731.00	31,748.30
Cempedak	7,675.40	5,568.60	34,741.00
Sapodila	542.5	472.9	5,356.40
Dokong	8,698.90	5,700.80	39,350.30
Duku	7,673.90	5,414.50	37,377.30
Durian	75,713.10	55,684.10	373,084.00
Guava	1,732.40	1,406.50	27,286.00
Langsat	5,451.80	3,358.30	21,488.90
Pomelo	1,092.80	888.20	11,714.20
Sweet Orange	3,107.60	1,988.20	18,665.90
Mango	5,270.40	3,259.80	16,625.00
Mangosteen	4,136.40	3,021.50	25,355.20
Dragon Fruit	611.60	451.80	8,577.00
Pineapple	10,580.30	8,543.80	244,353.00
Jackfruit	4,902.90	2,964.40	32,907.40
Banana	27,084.60	22,730.50	288,677.00
Pulasan	464.70	389.10	1,273.40
Rambutan	17,811.60	12,988.90	69,256.20
Salak	1,170.90	894.60	3,959.80
Watermelon	11,032.50	10,480.00	209,599.00
Total	197,533.6	148,644.1	1,513,178.00

Source: Department of Agriculture, Fruit crop statistics Malaysia 2013, DOA, Putrajaya, Malaysia, 2014a.

TABLE 3.10
Hectarage of Fruits in the Year 2011–2013

	Planted Area (ha)		
Item/Year	2011	2012	2013
Major fruits	219,870	199,586	197,533

Source: Department of Agriculture, Fruit crop statistics Malaysia 2013, DOA, Putrajaya, Malaysia, 2014a.

VEGETABLES AND CASH CROPS

Vegetables and cash crops are smallholders crops in Malaysia and occupy over 87,000 ha as shown in Table 3.11. The average size of farm is usually less than 1 ha. A few states in Malaysia, namely Pahang, Kelantan, Perak, and Johore are the places where the vegetables and cash crops production are concentrated. Cameron Highlands in Pahang has been traditionally used for growing temperate vegetables such as cabbage, lettuce, and tomatoes. Table 3.12 presents cabbage

TABLE 3.11
Hectarage and Production of Vegetables and Cash Crops in 2013

Category of Crops	Planted Area (ha)	Production (kg/ha)
Vegetables	67,777	1,434,200
Cash crops	19,269.86	237,892
Total	87,047	1,672,092

Source: Department of Agriculture, Vegetables and cash crops statistics, DOA, Putrajaya, Malaysia, 2014e.

TABLE 3.12
Hectarage and Production of Vegetables in 2013

Type of Vegetables	Planted Area (ha)	Production (Mt)
Asparagus	5.2	31.9
Spinach	4,287.60	56,649.00
Lady's finger	2,702.80	34,370.00
Broccoli	439.1	5,701.70
Chilli	4,104.40	59,775.00
Hot chilli	458.1	1,878.80
Sweet shoot	389.7	5,350.70
Spring onion	673.2	9,645.40
Celery	432.5	12,725.10
Chinese parsley	286	3,405.20
Vegetable corn	407.8	3,233.00
Chinese kale	887	13,531.00
Water spinach	3,775.70	43,900.70
Cabbage	4,844.70	129,148.00
Chinese cabbage	1,202.80	40,178.40
Chinese chives	117.5	2,354.60
Four-angled bean	247.2	2,087.90
French bean	964.8	10,884.00
Long bean	3,622.40	53,264.70

(*Continued*)

TABLE 3.12 (*Continued*)
Hectarage and Production of Vegetables in 2013

Type of Vegetables	Planted Area (ha)	Production (Mt)
Wax gourd	132.7	1,691.20
Snow pea	165.3	1,314.20
Butter bean	33.1	408.4
Sweet pea	20.2	181.4
Chinese box thorn	67	2,258.30
Cauliflower	1,011.60	8,984.70
Bottle gourd	614.2	11,956.60
Pumpkin	2,322.60	111,144.30
Carrot	76.4	1,272.90
Radish	412.5	8,196.30
Sweet pepper	1,009.70	60,719.20
Sweet leek	391.5	7,828.80
Maman	17.8	99.9
Meranti	19.7	122.9
Bitter gourd	1,360.40	17,428.50
Angled loofah	1,311.10	18,949.50
Spinach	540.2	7,994.70
Fern shoot	44.2	604
Pucuk keledek	16.5	242.7
Rebung buluh	22.2	15.5
Rebung madu	35.8	82
Remayong	3.8	75.6
Lettuce	3,345.70	64,993.30
Brassica	14,579.20	250,059.90
Watercress	210.2	4,389.00
Taugeh	0.2	3.2
Brinjal	2,173.70	54,234.30
Cucumber	5,160.70	119,857.10
Tomato	2,830.70	190,976.90
Total	67,777.4	1,434,200.40

Source: Department of Agriculture, Vegetables and cash crops statistics, DOA, Putrajaya, Malaysia, 2014e.

as the largest area occupying among vegetables cultivation, whereas sweet corn occupy the largest area in terms of cash crops as given in Table 3.13. Besides domestic consumption, Malaysia exports vegetables to other countries with the main market being Singapore. Due to high demand, area under vegetables and cash crops keep increasing tremendously over the years as shown in Table 3.14 for the year 2011–2013.

TABLE 3.13
Hectarage and Production of Cash Crops in 2013

Type of Cash Crops	Planted Area (ha)	Production (Mt)
Jagung (sweet corn)	9,719.80	86,499.00
Kacang Tanah (groundnut)	126.8	533.6
Ubi Kayu (cassava)	4,046.10	62,842.80
Ubi Keladi (yam)	405.3	2,934.10
Ubi Keledek (sweet potato)	3,041.60	50,747.50
Sengkuang (yam bean)	198.5	6,263.00
Tebu Kuning (sugarcane)	1,731.90	28,072.40
Jumlah (Total)	19,270.00	237,892.40

Source: Department of Agriculture, Vegetables and cash crops statistics, DOA, Putrajaya, Malaysia, 2014e.

TABLE 3.14
Hectarage of Vegetables and Cash Crops in 2011–2013

Item/Year	Planted Area (ha)		
	2011	2012	2013
Vegetables	35,955	47,750.70	67,777.4
Cash crops	15,509.80	16,156.50	19,270
Total	51,465	63,907.20	87,047.4

Source: Department of Agriculture, Vegetables and cash crops statistics, DOA, Putrajaya, Malaysia, 2014e.

FLORICULTURE

The floriculture industry has contributed significantly to the agricultural sector giving a net value of RM290 million in 2010, which is about 0.15% of GDP. This includes cut flowers, orchids, and nonorchids for the domestic and export markets. Total area involved with the floriculture industry in 2013 was 2234 ha. Total area under floriculture has increased steadily from 2213 ha in 2011 to 2227 ha in 2012 and 2234 ha in 2013 as given in Table 3.15. Production in floriculture is concentrated mainly in Johore and Pahang (Cameron Highlands). Area nearby the International Airport Subang, Sepang, Petaling, and Kuala Langat in the state of Selangor has been made as the third largest area for floriculture production.

LIVESTOCK

Livestock production in Malaysia can be characterized by two contrasting subsectors. The nonruminant sector, such as poultry and swine, is a highly commercialized

TABLE 3.15
Area under Floriculture in Malaysia

Item/Year	Planted Area (ha)		
	2011	2012	2013
Flowers	2213	2227	2234

Source: Ministry of Agriculture and Agro-Based Industry, Malaysia,
Agrofood statistics 2013. MOA, Putrajaya, Malaysia, 2014.

TABLE 3.16
Area under Livestock for 2008 and 2010

Livestock Area	Hectare (ha)	
Item/Year	2008	2010
Poultry, swine and others	5,796	6,366
Pasture/ruminant	17,368	17,506

Source: Ministry of Agriculture and Agro-Based Industry, Malaysia,
Agrofood statistics 2013. MOA, Putrajaya, Malaysia, 2014.

sector where large corporations are involved, and the total supply of the products is more than sufficient to meet domestic demand. In poultry production, high technology systems are being used, such as closed housing and computerized ration formulations and feeding method. The ruminant sector is mainly operated by smallholders and has shown little progress over the last decade with the self-sufficiency still very low. Table 3.16 shows that the landuse for nonruminant sector has increased more than ruminant sector in the past years.

AQUACULTURE

Aquaculture sector has proven to be an important supplier of animal protein. Aquaculture industry plays a vital role in providing social and economic stability to the industry players and the fisherman as a whole. Aquaculture activities such as farming of aquatic organisms, including fish, molluscs, crustaceans, and aquatic plants very contribute to the economy and the human life of the world. In Malaysia, aquaculture activities can be divided into two: freshwater and brackish water. Total landuse under aquaculture in Malaysia is over 38,000 ha as given in Table 3.17.

HERBS AND SPICES

A new subsector in Malaysian agriculture that is currently experiencing rapid growth is the herbs and spices industry. Hectarage and production of herbs and spices are

TABLE 3.17
Landuse for Aquaculture in 2013

Area for All Type of Aquaculture	Hectare (ha)
Freshwater	7,368.46
Brackish water	31,466.77
Total	38,835.23

Source: Ministry of Agriculture and Agro-Based Industry, Malaysia, Agrofood statistics 2013. MOA, Putrajaya, Malaysia, 2014.

given in Tables 3.18 and 3.19. Landuse under herbs and spices in Malaysia is over 7000 ha and seems to increase in the last few years as shown in Figure 3.4. The hectarage and types of herbs and production are given in Table 3.20, whereas the hectarage and types of spices and production are given in Table 3.21. The Landuse for herbs is dominated by petai, followed by jering and pandan as shown in Table 3.20, whereas the landuse for spices is dominated by lime, lemon grass, and ginger as shown in Table 3.21.

TABLE 3.18
Hectarage and Production of Herbs and Spices in 2013

Category of Crops	Planted Area (ha)	Production (Mt)
Herbs	1,298.87	8,424.78
Spices	5,859.64	52,255.61
Total	7,158.51	60,680.39

Source: Department of Agriculture, Herbs and spices statistics, DOA, Putrajaya, Malaysia, 2014b.

TABLE 3.19
Landuse under Herbs and Spices in 2011–2013

Item/Year	Planted Area (ha)		
	2011	2012	2013
Spices	4932.82	4879.52	5859.64
Herbs	1197.86	1041.19	1298.87
Total	6130.68	5920.71	7158.51

Source: Department of Agriculture, Herbs and spices statistics, DOA, Putrajaya, Malaysia, 2014b.

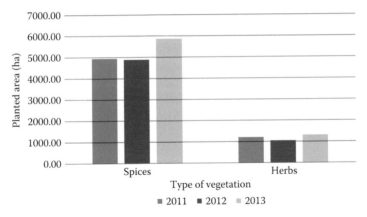

FIGURE 3.4 Land under herbs and spices in 2011–2013.

TABLE 3.20
Hectarage and Production of Herbs in 2013

Type of Herbs	Planted Area (ha)	Production (Mt)
Belalai Gajah (Belalai Gajah)	1	0.36
Cekur (Cekur)	6.27	20.88
Dukung Anak (Dukung Anak)	0.96	0.5
Gelenggang (Gelenggang)	1.4	0.28
Halia Bara (Ginger)	0.2	2.07
Jering (Jering)	114.01	469.42
Kaduk (Kaduk)	0.13	0.17
Kesum (Kesum)	19.38	592.16
Lidah Buaya (Aloe Vera)	41.21	467.46
Mas Cotek (Mas Cotek)	10	17.63
Melada Pahit (Melada Pahit)	3.2	0.1
Mengkudu/Noni (Great Morinda)	0.7	3.51
Misai Kucing (Misai Kucing)	9.7	47.73
Pandan (Pandan)	110.68	673.01
Pecah Beling (Pecah Beling)	0.4	0.25
Pegaga (Pegaga)	17.86	177.6
Pudina (Mint)	66.81	1537.75
Petai (Petai)	781.66	3495.46
Serai Wangi (Fragrant Lemon Grass)	36.73	212.15
Sirih (Betel Vine)	18.02	480.03
Stevia (Stevia)	0.4	0.07
Terung Pipit (Turkey Berry)	27.7	153.4
Tongkat Ali (Long Jack)	20.4	40.26
Ulam Raja (The King's Salad)	10.05	32.53
Jumlah (Total)	1298.87	8424.78

Source: Department of Agriculture, Herbs and spices statistics, DOA, Putrajaya, Malaysia, 2014b.

TABLE 3.21

Hectarage and Production of Spices in 2013

Type of Spices	Planted Area (ha)	Production (Mt)
Asam Gelugur (Asam Gelugur)	69.08	570.62
Asam Jawa (tamarind)	1.62	5.85
Bunga Kantan (pink cone ginger)	201.69	1,316.95
Cengkih (clove)	2.7	1.2
Halia (ginger)	892.68	8,830.68
Kari (curry)	23.94	48.45
Kayu Manis (cinnamon)	2.2	0.4
Kunyit (turmeric)	353.73	3,611.94
Lada Hitam (pepper)	2	2.3
Lengkuas (galangal)	355.99	3,119.15
Limau Kasturi (calamondin lime)	1,379.66	12,244.52
Limau Nipis (lime)	1,309.61	9,512.39
Limau Purut (kaffir lime)	40.41	246.04
Pala (Pala)	69.71	165.79
Selom (Selom)	7.03	39.03
Serai (lemongrass)	1,147.59	12,540.30
Total	5,859.64	52,255.61

Source: Department of Agriculture, Herbs and spices statistics, DOA, Putrajaya, Malaysia, 2014b.

CONCLUSION

Most of the arable land in Malaysia is dominated by plantation crops, particularly oil palm and rubber, occupying 84.25% of the total landuse in Malaysia. The next important crop is paddy, the staple food of Malaysia, which occupies 8.99%. The minor plantation crops that include coconut, coffee, tea, and others occupy 2.00%. The other crops including fruits, vegetables, herbs, spices, flowers occupy only 3.92%. The hectarage for aquaculture and livestock is very low, occupying only 0.84% of the total landuse in Malaysia.

Future landuse in Malaysia will remain to be dominated by the plantation crops, especially oil palm, followed by rubber. However, it is anticipated that with the current low prevailing price of commodity crops, especially rubber, the hectarage of plantation crop will be slightly reduced, and hectarage of other sectors especially fruits, herbs, and spices will be increased. As the percentage of landuse for aquaculture and livestock is very low, that is, 0.84%, it is pertinent that landuse for this sector should be increased, ensuring food security for Malaysia.

REFERENCES

Aminuddin, B. Y., W. T. Chow, and T. T. Ng, (1990). Resources and problems associated with sustainable development of upland areas in Malaysia. Blair, G. and Lefroy, R. (Eds.) In *Technologies for Sustainable Agriculture on Marginal Uplands in South-east Asia*, pp. 55–61. ACIAR Proceedings no. 33.

Arshad, F. M., Abdullah, R. M. N., Kaur, B., and Abdullah, M. A. (Eds.) (2007). *50 Years of Malaysian Agriculture: Transformational Issues, Challenges and Direction*. UPM Press, Serdang, Malaysia.

Department of Agriculture. (2014a). Fruit crop statistics Malaysia 2013. DOA, Putrajaya, Malaysia.

Department of Agriculture. (2014b). Herbs and spices statistics. DOA, Putrajaya, Malaysia.

Department of Agriculture. (2014c). Industrial crops statistics Malaysia 2013. DOA, Putrajaya, Malaysia.

Department of Agriculture. (2014d). Paddy statistics of Malaysia 2013. DOA, Putrajaya, Malaysia.

Department of Agriculture. (2014e). Vegetables and cash crops statistics. DOA, Putrajaya, Malaysia.

Malaysian Rubber Board. (2015). Natural Rubber Statistics. MRB, Kuala Lumpur, Malaysia.

Ministry of Agriculture and Agro-Based Industry, Malaysia. (2014). Agrofood statistics 2013. MOA, Putrajaya, Malaysia.

Noordin, W. D. (2013). *Rubber Plantation: Soil Management & Nutritional Requirement*. UPM Press, Serdang, Malaysia.

Olaniyi, A. O., Abdullah, A. M., Ramli, M. F., and Sood, A. M. (2013). Agricultural land use in Malaysia: An historical overview and implications for food security. *Bulgarian Journal of Agricultural Science*, 19, 60–69.

UPM. (2010). Agriculture and Man. UPM Press, Serdang, Malaysia.

4 Major Soil Type, Soil Classification, and Soil Maps

Aminaton Marto and
Safiah Yusmah Mohd Yusoff

CONTENTS

INTRODUCTION

Soil is a mixture of minerals, organic matter, gases, liquids, and some microorganisms. From the perspective of geotechnical engineering, the only important constituent of the soil is the mineral. Geotechnical engineering grouped the mineral in accordance with their compositional aspect, termed as gravel, sand, silt, and clay. Soils as one of the natural materials often exist as a mixture comprising two or more constituents. Malaysia is located on the tectonically stable plate of Sunda shelf. Malaysia consists of two regions: Peninsular Malaysia and East Malaysia. Half of the Peninsular Malaysia is covered by the igneous rock (granite); the rest are sedimentary rock (limestone); the remainder is alluvium as shown in Figure 4.1. Figure 4.2 shows the hydrogeological map of Sarawak, whereas Figure 4.3 shows the hydrogeological map of Sabah. This chapter encompasses the major soil types in Malaysia and their classification.

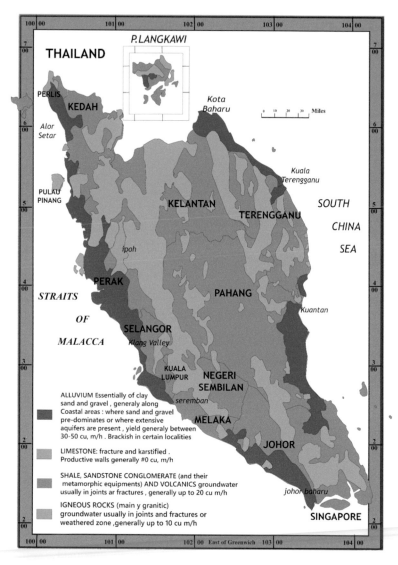

FIGURE 4.1 Simplified hydrogeological map of Peninsular Malaysia. (From Suratman, S., IWRM: Managing the groundwater component, in *Malaysia Water Forum*, Kuala Lumpur, Malaysia, 2004.)

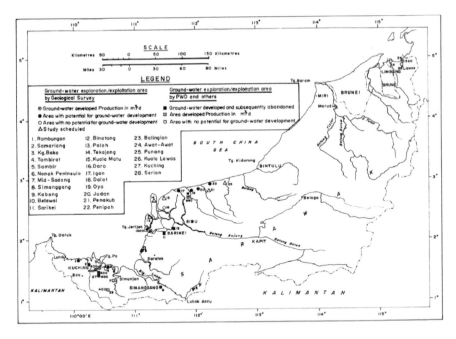

FIGURE 4.2 Simplified hydrogeological map of Sarawak. (From Suratman, S., IWRM: Managing the groundwater component, in *Malaysia Water Forum*, Kuala Lumpur, Malaysia, 2004.)

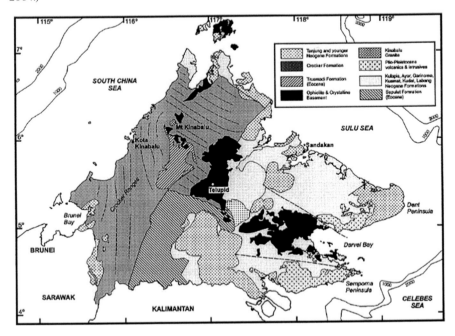

FIGURE 4.3 Simplified hydrogeological map of Sabah. (From Suratman, S., IWRM: Managing the groundwater component, in *Malaysia Water Forum*, Kuala Lumpur, Malaysia, 2004.)

MAJOR SOIL TYPES

Malaysia is a tropical climate country with hot and humid condition throughout the year; more than 70% of the land is covered by residual soil, whereas the remaining 30% is coastal alluvial soil. Parent rock type is the main geological factor in the formation of residual soils. Generally, there are three main soil groups in Malaysia: residual soils of granite, residual soils of sedimentary rock, and coastal alluvial soils; major soil groups are shown in Figure 4.4. Apart from that, about 3 million ha of the land in Malaysia is covered by peat. Hence, the following section discusses the three major soil types in Malaysia, namely the residual soils, alluvium, and peat. The soil maps for Peninsular Malaysia, Sabah, and Sarawak are shown in Figures 4.4 through 4.8.

RESIDUAL SOILS

Residual soils are formed from the weathering process of rocks. Therefore, the characteristics of the residual soil are typically similar to their parent rock. When the weathering rate exceeded the transportation rate of weathered soil, the soil remains in place as a residual soil. The intensive chemical weathering process in Malaysia causes the residual soil layers to be very thick, extending before reaching the unweathered parent rock as shown in Figure 4.9. The thickness of each weathering zones is variable. Granitic residual soil often extends more than 30 m, whereas residual soil of sedimentary rock is often thinner. Generally, the residual soils in tropical countries such as Malaysia always consist of layered soil profile that is parallel to the ground surface. The engineering properties of residual soils range from

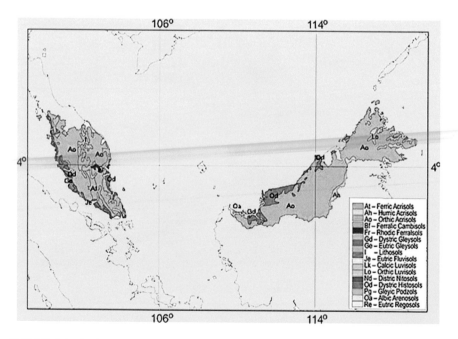

FIGURE 4.4 Major soils groups of Malaysia.

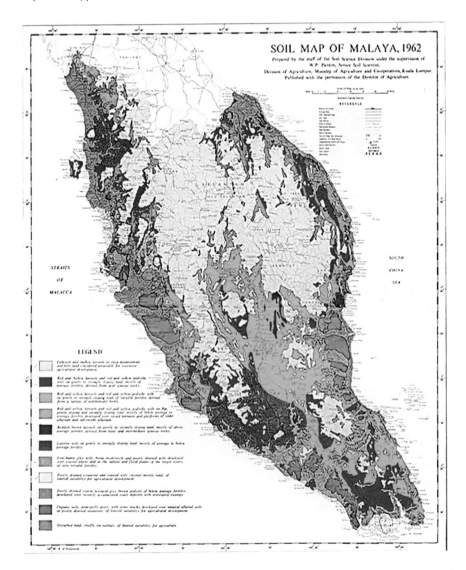

FIGURE 4.5 Soil map of Peninsular Malaysia. (From Ministry of Agriculture and Cooperatives, 1962.)

poor to good and generally improve with depth. The weathering profile reflects the progressive weathering process from the bedrock to the top soil. The weathering profile is subjected to the control of local variation such as rock type, climate change, topography, erosion rate, and others.

In general, Malaysia has a varied geology of igneous rocks, metamorphic rocks, and sedimentary rocks. Singh and Huat (2004) reported the aerial extent of each rock types by the surface area occupied by various lithological units as given in Table 4.1. The nature of the residual soils for each parent rock type had been explained by Tan et al. (2004).

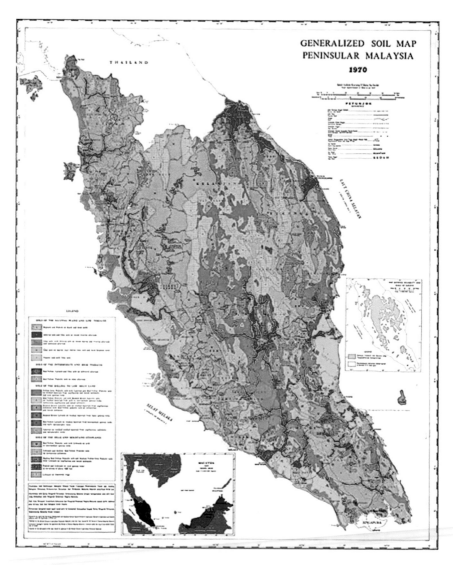

FIGURE 4.6 Soil map of Peninsular Malaysia. (From Ministry of Agriculture and Cooperatives, 1970.)

Granitic soils covered a large area of Peninsular Malaysia especially the hillside and mountain ranges. Granitic soils generally contain high sand content with low water content. Basaltic and andesitic soils are not as widespread as granitic soils. It can be found in sporadic patches of Peninsular Malaysia and large part of Tawau, Sabah. These residual soils are generally having higher water content because they contain more fine particles as compared to granitic soils. Gabbroic soils could be found in Johor, the southern part of Peninsular Malaysia. The properties are comparable to basaltic soils because the mineralogy of the parent rock type, for the basalt rock and gabbro rock, are similar.

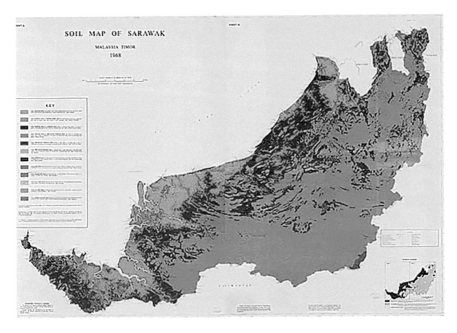

FIGURE 4.7   Soil map of Sarawak. (From Ministry of Agriculture and Cooperatives, 1968.)

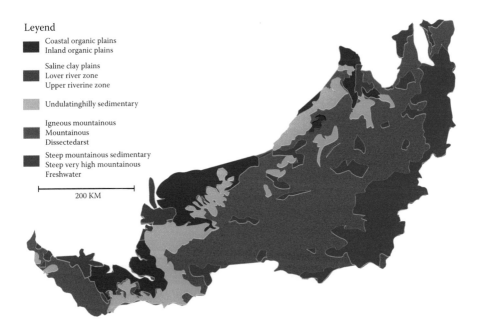

FIGURE 4.8   Soil map of Sabah. (From Ministry of Agriculture and Cooperatives, 1964.)

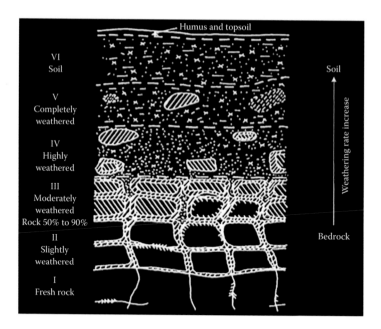

FIGURE 4.9 Typical classification of weathering profile. (From Little, A.L., The engineering classification of residual tropical soils, *Proceeding of Soil Mechanics & Foundation Engineering Conference*, Mexico, 1969.)

TABLE 4.1
Surface Areas of the Lithological Unit

Unit	Area (km²)
Igneous and metamorphic	48,319,327
Sedimentary—conglomerate	640,088
Sedimentary—limestone and marble	2,126,255
Sedimentary—sandstone	8,834,115
Sedimentary—shale, mudstone, slate	1,775,451
Quaternary	22,383,004
Undifferentiated lithology	47,883,066

Source: Singh, H. and Huat, B.B.K., Terra firma and Foundations, in *4th International Conference on Landslides, Slope Stability and Safety of Infrastructure.* March 24–25, Kuala Lumpur, Malaysia, 2004.

Shales and schists are the sedimentary rocks available in Malaysia. The carbonaceous shales and graphitic schists are often containing pyrite, which can cause further deterioration to the shear strength. The residual soils of both shales and schists are having low water content compared to quartz-mica schists. Residual soils from quartz mica are usually the silty soils type.

The residual soils of limestone are called residual red clay. The area covered by the residual red clay is known as slumped zone. Unlike the weathering profile of igneous rock, the residual soil of limestone is located immediately above the bedrock plane. The residual red clay is a very weak soil; it is very soft in terms of the strength. The slumped zone could be found in the central part of Peninsular Malaysia. It has been encountered at the site of some megaproject including Kuala Lumpur Convention Centre (KLCC) and Petronas Twin Tower (Mohamad et al. 1995).

ALLUVIAL SOILS

Alluvial soils are also known as fluvial soils or alluvium. These soils are transported to their present position by rivers and streams. When a river or stream is flowing rapidly, the silts and clays remain in suspension and are carried downstream. Only sand, gravels, and boulders will be deposited. However, when the water flows more slowly, more of the finer soils are also deposited. Therefore, alluvial soils often contain alternating horizontal layers of different soil types. These soils are very common all over the world, and many large engineering structures are built on them. In Malaysia, the alluvium is normally consolidated clay, exhibit consolidation due to desiccation and wreathing only. River in relatively flat terrain moves much slowly and often changes course, thus creating complex alluvial deposits (Indraratna et al. 1992). Figure 4.10 shows the area of Peninsular Malaysia covered by alluvium deposit.

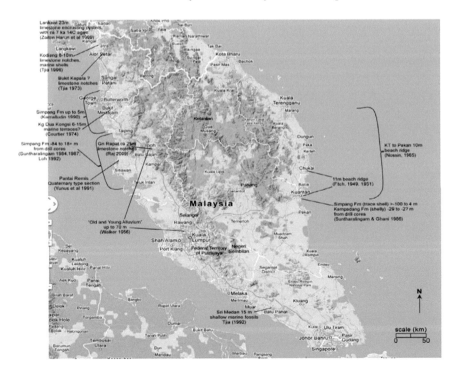

FIGURE 4.10 River and coastal alluvium in Peninsular Malaysia. (From Indraratna, B. et al., *J. Geotech. Eng.*, 118, 12–1, 1992.)

The alluvium plain in Sarawak extends along the shoreline area and almost parallel to the Rajang River through the division of Kuching, Sri Aman, Sarikei, and Sibu (Staub and Gastaldo 2003). Rajang River is the longest river in Malaysia. The place where the separation between the delta alluvium plain and the coastal alluvium plain occurs in Sarawak is at the midpoint between the Igan distributary of the Rajang River and the Oya River. Figure 4.11 shows the delta plain adjacent to the Rajang River, whereas Figure 4.12 shows the coastal plain adjacent to the Rajang River.

Peat

In general, soil with organic content greater than 20% is termed as organic soil where the mechanical criteria of conventional mineral soil can no longer be generally applied. The organic content is essentially the remains of plants whose rate of accumulation is faster than the rate of decay. The precise definition of peat, however, varies between the disciplines of soil science and engineering. Soil scientists define peat as soil with an organic content greater than 35%. However, geotechnical engineers term organic soil as soil with an organic content of more than 20%, whereas according to ASTM, 2013 peat is an organic soil with an organic content more than 75% (ASTM 2013). In Malaysia, USDA soil taxonomy was used in defining *peat soil* by the Department of Agriculture (DOA).

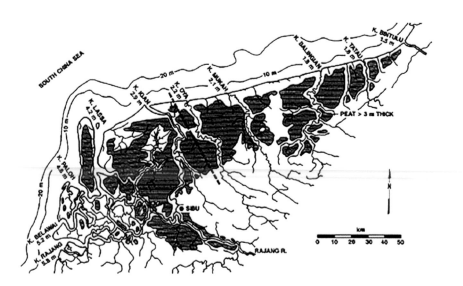

FIGURE 4.11 Delta plain adjacent to the Rajang River delta. (From Staub, J. R. and Gastaldo, R.A., Late Quaternary sedimentation and peat development in the Rajang River Delta, Sarawak, East Malaysia. Tropical Deltas of Southeast Asia Sedimentology, Stratigraphy, and Petroleum Geology SEPM Special Publication No. 76, 71–87, 2003.)

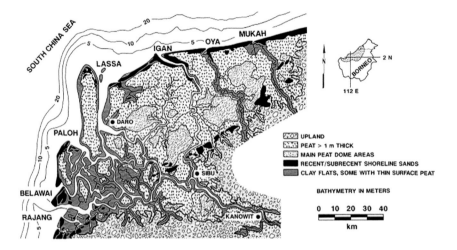

FIGURE 4.12 Coastal plain adjacent to the Rajang River delta. (From Staub, J. R. and Gastaldo, R.A., Late Quaternary sedimentation and peat development in the Rajang River Delta, Sarawak, East Malaysia. Tropical Deltas of Southeast Asia Sedimentology, Stratigraphy, and Petroleum Geology SEPM Special Publication No. 76, 71–87, 2003.)

According to the definition, peat is defined as soils, which have loss on ignition of more than 65%. Besides that, peat has also been classified in three categories by DOA based on depth: shallow peat (<150 cm), moderately deep peat (150–300 cm), and deep peat (>300 cm).

Peat commonly occurs as extremely soft, wet, and unconsolidated superficial deposits, normally as an integral part of wetland systems. Peat formation occurs along the inland edge of mangroves, where fine sediments and organic materials become trapped in the mangrove roots. It is a highly organic soil with an accumulation of disintegrated plant remains, which have been preserved under conditions of incomplete aeration and high water content. Therefore, the color of peat is usually dark brown or black with a distinctive odor. These characteristics also lend peat its own distinctive geotechnical properties compared to mineral soils, such as clay and sandy soils, which are made up only of soil particles. Peat is a very problematic soil and subjected to instabilities, discomfort, difficulty of access to the sites, and tremendous variability in material properties, and it can be a source of large differential settlements. However in Malaysia, peat is used widely for agriculture purposes, including oil palm (36%). According to Wetlands International (2010) report, the highest proportion of peat soil under agriculture is 44% for Peninsular Malaysia, whereas it is 33% for both Sabah and Sarawak.

In Malaysia, peat has been classified as one of the major soil group. Peat covers approximately 8% of the land or about 3 million ha as shown in Figure 4.13. Sarawak has the largest area of peat in the country, covering about 1.66 million ha

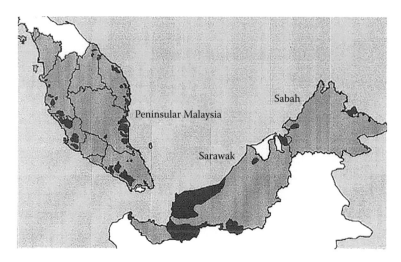

FIGURE 4.13 Distribution of peatland in Malaysia. (From Huat, B.B. et al., *Geotechnics of Organic Soils and Peat*, CRC Press, Boca Raton, FL, 2014.)

and constituting 13% of the state, followed by Peninsular Malaysia, which has about 984,500 ha, whereas Sabah has 86,000 ha. Only 20% of the peatlands in Malaysia remains as forest with canopy cover more than 70%. According to the report by Wetlands International (2010), the area covered by undisturbed peatlands is relatively similar for Peninsular Malaysia and Sarawak. In Peninsular Malaysia, Pahang has the largest area of good quality peat swamp forest (129,759 ha), followed by Terengganu (51,759 ha). Overall, Sarawak has the largest area of peat soils (1,697,847 ha), followed by Peninsular Malaysia (642,918 ha) and Sabah (116,965 ha) (Wetlands International 2010).

The report from Wetlands International (2010) also mentioned that more than 510,000 ha of peatlands in Malaysia are planted with oil palms with the largest in Sarawak (300,000 ha in 2008). Under agriculture, Johor has the greatest area, followed by Perak and Selangor. In Peninsular Malaysia, about 281,652 ha of peat soils are under cultivation, and 72% are under oil palm plantations. The greatest extent of oil palm on peat soils is Johor (68,468 ha) followed by Perak (61,257 ha). However, with 88%, Perak has the highest proportion of its peatlands under oil palm, compared to about 48% in Johor (Wetlands International 2010). Paramananthan (2000) listed the peat soil distribution and the area developed for agriculture in Malaysia, as shown in Table 4.2.

Peat formation in Selangor is found at the river basins between Selangor River and Langat River. According to Wetlands International (2010), about 121,276 ha of total peat swamp forest area in Selangor was untouched in the 1950s. Nowadays most of the areas have been cleared for agriculture. Peat area has a great threat of forest fires. It has been reported that about 500 ha was destroyed by fire in 1998. In fact, some areas dominated by herbaceous vegetation with little tree regeneration experienced frequent fires (Wetlands International 2010).

TABLE 4.2
Peatlands Distribution in Malaysia

Region	State/Division	Extent of Organic Soils (ha)	Area Developed for Agriculture (ha)
Peninsular Malaysia	Johor	205,856	145,900
	Kelantan	7,880	2,100
	Negeri Sembilan	8,188	5,000
	Pahang	228.644	17,100
	Perak	74,075	69,700
	Selangor	186,602	65,900
	Terengganu	85,537	13,900
	Subtotal	796,782	313,600
Sabah	Subtotal	200,600	Na
Sarawak	Kuching	26,827	Na
	Samarahan	205,479	50,836
	Sri Aman	340,374	50,836
	Sarikei	172,353	61,112
	Sibu	502,466	269,571
	Bintulu	168,733	47,591
	Miri	314,585	66,114
	Limbang	34,730	8,715
	Subtotal	1,765,647	554,775
Malaysia total		2,762,929	868,375

Source: Paramananthan, S., *Soils of Malaysia: Their Characteristics and Identification*, Vol. 1, Academy of Sciences Malaysia, Kuala Lumpur, Malaysia, 2000.

In Johor, about 6300 ha of the peatlands is found in Pontian, Batu Pahat, and Muar in West Johor (Yulindasari 2006). Peat soils are found on marine soils, acid sulfate soils, and marine clays. At Johor west coast area, peat is found overlying acid sulfate soil, whereas at the east coast the peat overlies sand and clay. The presence of acid sulfate subsoils may pose a serious problem to cultivated areas of peatland, particularly along the west coast of Johor. The acid sulfate subsoils may eventually surface with the loss of peat soil, which caused problems as crop hardly survives on acid sulfate soils. Besides that, the environment may deteriorate to a level that local communities will have to abandon the area. This is because acid sulfate soils release large quantities of acidic water to adjoining canals, especially during the rainy seasons. If this happens, the water cannot be used for domestic purposes, will cause severe corrosion of metal structures, and will not be suitable for fish breeding. As pointed by Wetlands International (2010), only few species of fish may survive under these conditions.

Pahang has a total of 197,768 ha of peatlands that includes Pahang River North Peatland as well as the Pekan, Nenasi, Kedondong, and Resak Forest Reserves. Pekan Forest Reserve has the largest area of peatland (59,097 ha), whereas Nenasi

Forest Reserve has an area of 20,546 ha. Others like Resak and Kedondong Forest Reserves have smaller blocks of peatlands (9681 ha and 1818 ha, respectively). In Terengganu, fauna and many bird and reptile species were once the important habitats of the peat swamps before the area was developed. For Perak state, it originally had an area of 70,418 ha of peat swamp forest, which was located mainly along the coastal areas of the state. The peat swamp forests helped to mitigate the flood and to regulate storm damage. For many years, scientific-based research has been conducted at Beriah peat swamp forest, located at the north of Perak. It was found that this peat swamp forest supplies water for the nearby Keriah rice fields (Wetlands International 2010).

Kelantan has peat swamp forest that covers approximately 9146 ha: Berangan peatland (1392 ha), Pasir Puteh peatland (6433 ha), and Pasir Mas peatland (1311 ha). When paddy was extensively grown as the main food crop in the 1970s, the peat swamp was cleared. About 605 ha of peatlands was destroyed by forest fires between 1998 and 2000. According to the report by Wetlands International (2010), *Gelam* trees grow extensively in these burnt and degraded areas.

In Negeri Sembilan, there are five small blocks of peatland areas found at the coastal region and a sixth block at the inland close to the Johor State boundary. Covering a total area of 6245 ha, 68% of the peatlands are used as agriculture land, including oil palm, rubber, mixed horticulture, and diversified crops (Wetlands International 2010). In Selangor, about 380 ha of the Kuala Langat North Peatlands is located within the Putrajaya, including 44 ha of relatively undisturbed peat swamp forest. Twenty-four percent of the peatlands are under infrastructure for the new townships, whereas 21% are severely degraded peatland (Wetlands International 2010).

There are about 116,965 ha of peat soils in Sabah in which most of the soils are in Klias Peninsula and Kinabatangan–Segama Valleys. Previously before the conversion to agriculture land, the Klias Peninsula had about 60,500 ha of peat swamp forest. For Kinabatangan Valley, the largest is at the Lower Kinabatangan–Segama Wetlands Ramsar Site, which includes 17,155 ha of intact peatland forest from a total of 78,803 ha of coastal wetland ecosystems. Others are at the Kinabatangan Valley between Batu Puteh and Bilit, covering about 5000 ha. As reported by Wetlands International (2010), much of the area has been badly affected by fire in the west close to Batu Puteh.

In Malaysia, peat is usually found on the inward edge of mangrove swamps along a coastal plain and termed as either basin or valley peat (Huat et al. 2014). The depth of the peat is generally shallower near the coast and increases inland, locally exceeding more than 20 m. Many peatlands are far inland developed along the former coastline. For example, in Marudi, Sarawak, the peatland covered about 100 km inland. Water plays a fundamental role in the development and the maintenance of tropical peat. Rainfall and surface topography regulate the overall hydrological characteristics of peatland. Peatland is also generally known as wetland or peat swamp because of its water table, which is close to or above the peat surface throughout the year and fluctuates with the intensity and frequency of rainfall. About 60% of the peatland could be found in Sarawak as shown in Figure 4.14. Out of the 6500 km^2 delta plain of Rajang River, about half of it is covered by peat of greater than 3 m thick. It is also discovered that the peat deposits in northeast of the Daro town extend laterally. As for the alluvial valley of the coastal plain, peat

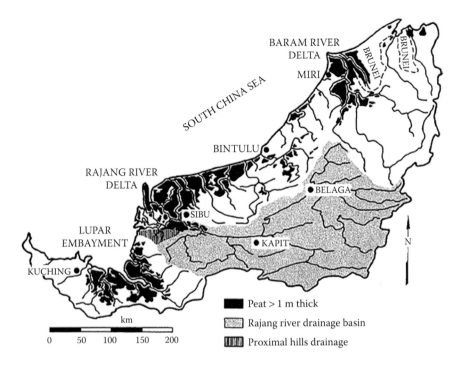

FIGURE 4.14   The peatland distribution in Sarawak. (From Staub, J.R. and Gastaldo, R.A., Late Quaternary sedimentation and peat development in the Rajang River Delta, Sarawak, East Malaysia. Tropical Deltas of Southeast Asia Sedimentology, Stratigraphy, and Petroleum Geology SEPM Special Publication No. 76, 71–87, 2003.)

deposits are located between the towns of Sibu and Kanowit (Staub and Gastaldo 2003) as shown in Figure 4.14. It covers 400 km^2 with maximum reported peat thickness of more than 20 m.

The report from Wetlands International (2010) says that the largest area of peat swamp forest in Malaysia was in Sarawak. However, due to logging activities the peat swamp forest area remained only about 1,054,844 ha in 2000 with less than 20% of this is classified as undisturbed peat swamp forest. A total area of 18,920 ha out of the total of 223,277 ha for undisturbed forest was considered to be in pristine area that confined to Miri and Limbang Divisions. Limbang Division has a total area of 34,730 ha of peat soil; Miri Division has 295,995 ha, whereas Bintulu Division has a total peatland area of 157,442 ha. A total of 340,374 ha of the peatland area is covered by Sri Aman and Betong Divisions. Sarikei Division has a peatland area of 74,414 ha. Total area under crop cultivation is 61,112 ha. Sibu and Mukah Divisions have a total peatland area of 600,387 ha, but the area has been subjected to intensive logging since 1961. Besides that, the oil palm plantation covers many parts of the original peat swamp forest. For Samarahan Division, the total area of peatland is 165,581 ha, whereas in Kuching Division it is 26,827 ha. The peat soil area not covered by forest has been converted to agriculture or infrastructure as a result of economic development.

Besides being used widely for agriculture purposes, Huat et al. (2014) reported that the peatlands of Malaysia play an important role in regulating and reducing flood damage, preserving water supply, providing fish, timber, and other resources for local communities, besides regulating the release of greenhouse gases by storing large amounts of carbon within peat. Hence, the Malaysian government is working hard to minimize the clearance of peat swamp forest, although it is necessary for economic development. The respective authorities are now taking proactive plan and strategies to preserve the area, which now becomes a tourist attraction in certain part of the country.

SOIL CLASSIFICATION

Soil classification deals with the systematic categorization of soils based on distinguishing characteristics as well as criteria that dictate choices in use. Peninsular Malaysia, Sabah, and Sarawak have used different criteria for the identification technique to produce the soil map. The use of different definitions of soil series and soil classification in these three regions has been mainly due to historical reasons (Paramananthan 2000). No proper classification has been fixed for Peninsular Malaysia; the Soil Survey Division of Department of Agriculture had defined about 290 soil series based on the *Keys to Soil Taxonomy* (Soil Survey Staff 1994). As for in Sabah, soil families were defined using the older draft versions of *UNESCO Soil Map of the World Legend* (FAO 1988). A total of 107 soil families have been identified. In Sarawak, reconnaissance soil surveys of the different regions were carried out by different surveyors. Therefore, different soil classification has been used to interpret their results. Teng (2004) summarized these earlier soil classification systems and produced the *Keys to Soil Classification in Sarawak*. About 140 soil series have been identified for Sarawak under this classification system. There is not much attempt to evolve a common classification system for the whole Malaysia.

Paramananthan (1978) initiated the *Proposals for a Unified Classification of Organic Soils of Malaysia* to correlate some common soils of Peninsular Malaysia, Sabah, and Sarawak with the collaboration effort from the Department of Agriculture (DOA) Malaysia. However, still no soil map for the whole of Malaysia has been published based on this classification system.

From geotechnical engineering perspective, soils are classified according to their use for foundation support or building materials. The most common engineering classification system for soils is the Unified Soil Classification System (USCS). The USCS has three major classification groups: (1) coarse-grained soils (e.g., sands and gravels), (2) fine-grained soils (e.g., silts and clays), and (3) highly organic soils (referred to as *peat*). The USCS further subdivides the three major soil classes for clarification. Other popularly used engineering soil classification systems include the American Association of State Highway and Transportation Officials (AASHTO) System. A full geotechnical engineering soil description in AASHTO will also include other properties of the soil including color, in situ moisture content, in situ strength, and more detail about the material properties of the soil than is provided by the USCS. However, the AASHTO classification system is mainly used for disturbed soils, as in the case of highway subgrades. It could not be used for classifying soils in designing a foundation for structures (Das 2004). On the other hand, there are also researchers who used British Soil Classification System (BSCS) to classify the soils particularly for engineering purposes. The classification is quite similar with the USCS. The following sections discuss the classification of the three main soil types in Malaysia.

RESIDUAL SOILS

There are lots of classification systems of residual soils available, but there is still no universal scheme that could be practically applicable worldwide. Generally, the classification is based on either their weathering profile or the pedological groups. The classification based on weathering profile proposed by Little (1969) is the most commonly used, and it is practical to be used for those igneous rocks in humid tropical climate such as in Malaysia. Figure 4.15 shows the variations in residual soil weathering profile as proposed by Little (1969). Besides that, classification could be done according to the pedological terms.

Laterite is one of the most common types of residual soils found in tropical regions including Malaysia. It is one of the granitic residual soils that derived from igneous rock. Laterite is rusty red in color due to the presence of iron oxides. The basic characteristic of laterite can be varying significantly, depending on the origin, climate, and depth. The mineralogy of laterite found in Malaysia consists of two main minerals, which are quartz and kaolinite, whereas montmorillonite, illite, and muscovite are the minor minerals (Marto and Kasim 2003).

In general, Malaysia has a varied geology of igneous rocks, metamorphic rocks, and sedimentary rocks. The characteristic of residual soil is largely dependent on their parent rock. Table 4.3 lists the classification of residual soils obtained from various locations in Malaysia.

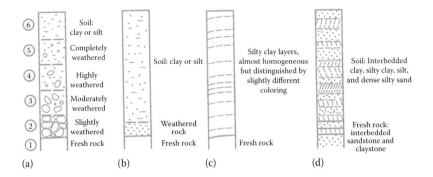

FIGURE 4.15 Variations of residual soil weathering profiles. (a) Gradual weathering profile–typical of weathered granite; (b) Sharp transition from rock to soil–typical of weathered basalt; (c) Uniform layers, degree of weathering not necessarily related to depth–typical of volcanic ash; and (d) Stratified nature of parent rock reflected in soil profile–typical of weathering of soft sedimentary rock, especially sandstone. (From Wesley, L.D. *Geotechnical Engineering in Residual Soils*, John Wiley & Sons, Hoboken, NJ, 2010.)

TABLE 4.3
Classification of Some Residual Soils in Malaysia

Location		Parent Material	Classification	Source
Johor	Amber Estate	Granodiorite	Yellowish brown clay	Pushparajah and
	Ayer Hitam	Shale	Light brownish grey clay	Amin (1977)
	Kulai	Rhyolite	Yellowish brown clay	
	Muar	Andesite	Reddish brown clay	
	Senai	Cabbro	Brown clay	
	Sedenak	Granite	Material Handling Systems (MHS)	Marto and Kasim (2003)
	Senai		CLAY of extremely high plasticity (CE)	
	Mutiara Rini		Chlorhexidine Gluconate (CHG)	
	Segamat	Andesite	Reddish clay	Maail et al. (2004)
	Skudai		Micro High (MH)	Salih et al (2012)
Kelantan	Batang Merbau	Schists	Yellowish brown sandy clay	Pushparajah and Amin (1977)
	Pulai Chondong	Granite	Clay of High (CH)	Marto and Kasim (2003)
Selangor	Serdang	Shale	Yellowish red clay	Maail et al. (2004)
		Sandstone	Bownish sandy clay	
		Granite	Reddish clay	
		Schist	Yellowish red sandy clay	
Kuala Lumpur	Cheras		CH	Taha et al. (2002)
			CH	Salih et al. (2012)
	Central		Soil Morphology (SM)	
	Bangsar (Umbrio soils [UM])		Micro Low (ML)	
Melaka	Pagoh	Shale	Brown clay	Pushparajah and Amin (1977)
	Bandar Melaka		MH, CH	Kepli (1994)
			MH	Salih et al. (2012)
Negeri Sembilan	Nilai		Clayey silt, Sandy silt	Anuar and Faisal (1997)
	Tampin	Granite	Consumer Value Store (CVS), Sydney Morning Herald's (SMH)	Marto and Kasim (2003)
Pahang	Bukit Goh	Basalt	Dark yellowish brown clay	Pushparajah and Amin (1977)
	Karak		Microslope (ML)-Micro High (MH)	
	Kuantan	Shale	Brown clay	
	Pekan		Dark brown sandy clay	
	Temerloh		Reddish brown sandy clay	

(Continued)

TABLE 4.3 (*Continued*)
Classification of Some Residual Soils in Malaysia

Location		Parent Material	Classification	Source
	Brinchang	Granite	Clay Low sands (CLS), Very silty SAND (plasticity sub-group) [SML]	Marto and Kasim (2003)
	Kuantan	Basalt	Dark brown clay	Maail et al. (2004)
	Lanchang	Granodiorite	Brownish clay	
Penang	Bukit Mertajam	Granite	Crops Horticulture and Soils (CHS), Schedule (SCH)	Marto and Kasim (2003)
	BatuFerringi	Granite	Sandy silt	Ahmad et al.
	PayaTerubong		Sandy silt	(2006)
	Bukit Bendera		ML	Salih et al. (2012)
	Sungai Ara		CL	
	TanjungBungah		Sandy silt	
Perak	Kledang		Clay low-Clay high(CL-CH)	Salih et al. (2012)
Selangor	Sungai Buloh	Granite	Yellowish brown sandy clay	Pushparajah and
		Sandstone	Dark brown sand	Amin (1977)
		Shale	Brownish yellow clay	
Terengganu	Dungun	Shale	Dark greyish brown fine sand	Pushparajah and Amin (1977)
	WakafTapai	Granite	CLAY of very high plasticity (CV), CVS	Marto and Kasim (2003)
	Bukit Temiang	Granite	Reddish yellow clayey sand	Maail et al. (2004)

Note: Marto and Kasim (2003) used the British Soil Classification System (BSCS).

ALLUVIAL SOILS

According to Paramananthan (2012), there are six types of the alluvial soils, developed over alluvial deposits. They are as follows:

1. *Older alluvium*: This type of alluvial soils normally occurs on gently rolling or at elevations of about 30 m. Valleys between these low hills are *U* shaped and broad. Colors in these soils get paler with depth, and rounded water-worn pebbles (parent material) occur at different depths. These soils can be confused with soils over conglomerate, but in these soils color becomes redder with depth and the valleys become *V* shaped.
2. *Subrecent nonaccreting alluvium and colluvium*: These soils normally occur on gently undulating to level terrain at elevations of 15–30 m. Termite mounds are common on such terrain. These terraces are associated with old rivers and mostly occur away from large rivers. Red mottles

are common in these soils due to a fluctuating water table. Drainage classes and textural classes are used to separate these soils.

3. *Recent accreting alluvium*: These are soils formed on recent floodplains of the larger rivers. Depending on the surrounding hills, mica flakes may be present. Lithologic discontinuities are common below 50 cm depth. Manganese nodules or specks are common in these soils. They occur on level terrain and are widely used for wetland rice cultivation.

4. *Beach deposits*: These are deposits often referred to as BRIS soils—beach ridges interspersed with swales. The sandy deposits often form ridges, whereas the swales are variable ranging from sand, clay, and even organic.

5. *Sulfidic materials/sulfuric horizon*: Marine or brackish water sediments (clays or sands rich in pyrite [Fe_2S]) are characterized by their hydrogen sulfide (rotten eggs) smell. When these are drained, the sulfides are oxidized to sulfuric acid, yellow jarosite mottles appear, and the soil pH drops to <3.5 (sulfuric horizon). To be significant, the sulfidic materials or the sulfuric horizon must occur within 100 cm of the soil surface.

6. *Marine alluvium, estuarine, and brackish water deposits*: These are soils formed on nonsulfidic deposits near the coast or adjacent to peat swamps. Similar to sulfidic materials, they were waterlogged but are now drained for cultivation.

Among all the six types of alluvial soils, the most highly discussed by the geotechnical engineer is the marine alluvium, in particular the marine clay. This is because this type of soil has low shear strength and high compressibility. Before any structure could be constructed on this type of soil, a proper soil improvement technique has to be carried out first in order to improve the properties so that it will not be detrimental to the construction. Generally clay is a fine-grained soil that combines one or more clay minerals with traces of metal oxides and organic matter. Geologic clay deposits are mostly composed of phyllosilicate minerals containing variable amounts of water trapped in the mineral structure. Clays are distinguished from other fine-grained soils by differences in size and mineralogy. Silts, which are fine-grained soils that do not include clay minerals, tend to have larger particle size than clays. There is, however, some overlap in particle size and other physical properties, and many naturally occurring deposits include both silt and clay. The distinction between silt and clay varies by discipline. Geologists and soil scientists usually consider the separation to occur at a particle size of 2 μm (clays being finer than silts), sedimentologists often use 4–5 μm, and colloid chemists use 1 μm (Pansu and Gautheyrou 2007). Geotechnical engineers distinguish between silts and clays based on the plasticity properties of the soil, as measured by the soils' Atterberg limits as well as based on particles as being smaller than 2 μm.

Marine clay deposits found at the coastal region of Malaysia are generally formed as the result of a secondary sedimentary deposition process after they have been eroded and transported from their original location of formation. There are three classification of marine clays found in Malaysia: kaolinite, montmorillonite–smectite, and illite. Naturally, these marine clays are mixtures of these different types, along with other weathered minerals. Table 4.4 shows the classification of some marine clays obtained from various locations in Malaysia.

TABLE 4.4

Classification of Some Marine Clays in Malaysia

Location		Classification	Source
Johor	DesaTebrau	Yellowish sandy clay (older alluvium)	Mohamad et al. (2011)
	Kota Tinggi	Dark yellowish brown sandy clay (older alluvium)	Pushparajah and Amin (1977)
	Sungai Tiram	Dark greyish brown sandy loam (older alluvium)	
	UluTiram	Dark greyish brown sandy loam (older alluvium)	
	Nusajaya	Dark greyish clay (older alluvium)	Marto et al. (2015)
Kedah	Kulim	Dark greyish brown sandy loam (riverine alluvium)	Pushparajah and Amin (1977)
Perak	Dinding	Dark greyish brown sandy loam (riverine alluvium)	Pushparajah and Amin (1977)
	Kinta	Dark greyish brown sandy loam (riverine alluvium)	
	Krian	Dark greyish brown mucky clay (marine alluvium)	
		Pale brown silty clay loam (alluvial plain)	
	Setiawan	Greyish brown clay loam (riverine alluvium)	
Selangor	Sungai Tinggi	Dark greyish brown silty clay (marine alluvium)	Pushparajah and Amin (1977)
	Sungai Buloh	Dark yellowish brown sandy loam (riverine alluvium)	

PEAT

Generally, the physical, chemical, and physicochemical properties of peat such as texture, organic content, mineral content, pH, color, water content, and degree of decomposition are served as a basis for its classification. Other classifications are based on the origin of peat and the field conditions during deposition (Huat et al. 2011). Most of the time, peats are classified on the basis of the constituent plants rather being based on their texture and composition. Decomposition is the breakdown process of plant remains by the soil microflora, bacteria, and fungi via aerobic decay. The degree of decomposition depends on a combination of conditions including the chemistry of water supply, temperature of the region, aeration, and the biochemical stability of the peat-forming plant.

The degree of humification could be obtained by performing the von Post test. This is done by taking a handful of peat and squeezing it on the palm of the hand. The liquid or other material extruded between the fingers and the residue remaining in the hand were examined and compared with the von Post scale shown in Table 4.5. In this classification system, there are 10 degrees of decomposition ranging from H1 (very fibrous) to H10 (very few fibers), which represent the state of decomposition/decay of the organic plant remains. The higher the number in von Post scale, the higher the degree of decomposition. Usually, peats near the surface fall into the H3 and H4 categories, but with increasing depth it would be classified as H5–H7. Huat et al. (2014) further classified the peats in Malaysia into three groups as summarized in Table 4.6.

Peat could also be classified in accordance to the fiber content as given in Table 4.7. The fiber content is determined from the dry weight of fiber retained on a #100 sieve (>0.15 mm opening size) as the percentages of oven-dried mass. Fibric peat is mostly undecomposed, extremely acidic, and typically yellowish

TABLE 4.5
Peat Classification on von Post Scale

Degree of Humification	Decomposition	Description	Material Extruded between Fingers	Residual in Hand
H1	None	Entirely unconverted mud-free peat	Clear, colorless water	
H2	Insignificant	Almost entirely unconverted mud-free peat	Yellowish water	
H3	Very slight	Very slightly converted of very slightly muddy peat	Brown, muddy water, no peat	Not pasty
H4	Slight	Slightly converted of somewhat slightly muddy peat	Dark brown, muddy water, no peat	Somewhat pasty
H5	Moderate	Fairly converted or rather muddy peat, plant structure still quite evident	Muddy water and some peat	Thick, pasty
H6	Moderate strong	Fairly converted or rather muddy peat, plant structure indistinct but more obvious after squeezing	About one third of peat squeezed out, water dark brown	Very thick
H7	Strong	Fairly converted or markedly muddy peat, plant structure still dissemble	About one half of peat squeezed out, consistency like porridge, any water is very dark brown	
H8	Very strong	Well converted or very muddy peat, very indistinct plant structure	About two thirds of peat squeezed out, also some pasty water	Plant roots and fibers, which resist decomposition
H9	Nearly complete	Almost completely converted or very mud-like peat, plant structure almost not recognizable	Nearly all the peat squeezed out as a fairly uniform paste	
H10	Complete	Completely converted or entirely muddy peat, no plant structure visible	All the peat passes between the fingers, no free water visible	

Source: Huat, B.B. et al., *Geotechnics of Organic Soils and Peat*, CRC Press, Boca Raton, FL, 2014.

TABLE 4.6
Three Classification Groups of Peat

Designation	Group	Description
Fibrous	H1–H4	Low degree of decomposition, fibrous structure, easily recognized plant structure, primarily of white masses
Pseudofibrous	H5–H7	Intermediate degree of decomposition, recognizable plant structure
Amorphous	H8–H10	High degree of decomposition, no visible plant structure, Mushy consistency

Source: Huat, B.B. et al., *Geotechnics of Organic Soils and Peat*, CRC Press, Boca Raton, FL, 2014.

TABLE 4.7
Classification of Peat Based on Fiber Content

Categories	Fiber Content (%)	Von Post Scale	Description
Fibric	>67	<H4	Fibrous, least decomposed
Hemic	33–67	H5/H6	Semifibrous, intermediate decomposed
Sapric	<33	H7	Amorphous, most decomposed

Source: Huat, B.B. et al., *Geotechnics of Organic Soils and Peat*, CRC Press, Boca Raton, FL, 2014.

TABLE 4.8
Classification of Some Peats in Malaysia

Region	Location	Organic Content	Classification	Sources
Peninsular	Johor	80–96	Hemic	Zainorabidin and Wijeyesekera (2007)
Malaysia	Klang, Selangor	78–85	H3–H6	Huat (2006)
	Banting, Selangor	83–94	H1–H6	
	Negeri Sembilan	83–98	H2–H5	
	West coast	31–77	H4–H8	Duraisamy et al. (2007)
	TasikBera	65–80	Fibric	Wust and Bustin (2004)
Sarawak	Sibu	25	H4 Sapric	Kolay et al. (2011)
	Mukah	45–48	Fibric	Melling et al. (2005)
	Kota Samarahan	30	Sapric	Fong and Mohamed (2007)
Sabah	Kinabatangan	50–95	Fibrous	Zainorabidin and Wijeyesekera (2007)

brown to light reddish brown in color. It has high organic content, with low degree of humification. Hemic peat is typically dark reddish brown in color, whereas sapric peat is darker: dark grey or black in color and is quite stable in its physical properties. Sapric peat contains highly decomposed organic matters with less fiber, and the water-holding capability is generally the lowest. With lowest void ratio, the permeability of sapric peat is also lowest, with lower compressibility and friction angle. Table 4.8 shows the classification of some peats found in Malaysia.

AGROECOLOGICAL ZONES

Malaysia is divided into two geographical regions by the South China Sea. These two regions, namely West or Peninsular Malaysia and East Malaysia resulted in a total land area of 339,733 km^2. Although both regions are under the same warm humid tropics, their climatic and agroecological environments are slightly different; hence, they are described separately.

Ultisols and Oxisols (classified under USDA Soil Taxonomy), which are acidic and highly weathered, formed 72% of Malaysian soils. Both of these soils have effects on phosphate fertilizer uses due to fixation. However, by adding 2–4 tonnes/ha of limestone periodically, the problem is solved. Except for certain highlands such as Cameron Highlands and Pahang, other main regions experience similar rainfall and thermal patterns, as well as lengths of growing period. In the highland areas, crops that suit the cool environment such as tea and strawberry are grown. Malaysia does not use agroecological zones (AEZs) to plan for crop production. Instead, they are based on state or regional (west or east Malaysia) classifications. Various kinds of crops are grown in Malaysia but the main crops differ from state to state in terms of area and production depending on the crops' suitability to local weather and soil conditions. The rank of crops suitability is shown in Table 4.9 for all the states in Malaysia.

PENINSULAR MALAYSIA/WEST MALAYSIA

Peninsular Malaysia or West Malaysia, which consists of 12 administrative states, has an area of about 132,000 km^2 (13 million ha). The states in Peninsular Malaysia are Perlis, Kedah, Pulau Pinang, Perak, Selangor, Wilayah Persekutuan, Negeri Sembilan, Melaka, Johor, Pahang, Terengganu, and Kelantan. Eighty percent of Peninsular Malaysia is lowland, and the remainder is highland. More than 60% of the area is arable and cultivated with soil and ecological suitability crops. These crops are cultivated in the AEZs in Peninsular Malaysia as shown in Figure 4.16.

In order to improve fertilizer management and application efficiency, knowledge of the soils is important. The type of soil will determine which fertilizers are to be applied. There are three groups of lowland soils in Malaysia, which are considered as problematic soils: BRIS/sandy soils (beach ridge [Bris] and tin tailings), peat soils, and acid sulfate soils. To ensure high success rate of crop cultivation, these soils need to be enriched. About 155,400 and 40,400 ha of BRIS soils exist in Peninsular Malaysia and Sabah, respectively. The total area of tin-tailing areas in Peninsular Malaysia in 1995 was estimated at 200,000 ha and were expanding at a rate of 4,000 ha annually. The serious problems faced in sandy and peat soils are related to the loss of applied fertilizer by leaching. Therefore, farmers have to adopt suitable farming techniques and management practices, for instance, application of fertilizers depends on specific crop requirements. As for acid soils, phosphate fixation is the main problem in the context of fertilizer application. However, this problem is easily dealt by using lime amendment and direct application of phosphate rocks.

TABLE 4.9

Recommendation of Suitable Crops for Cultivation in Particular States in Malaysia

STATES

CROPS	Johor	Kedah	Kelantan	Melaka	Negeri Sembilan	Pahang	Perak	Perlis	Pinang	Sabah	Sarawak	Selangor	Terengganu
Rubber	S	L	S	S	S	S	S	L	S	S	S	S	S
Coconut	S	S	S	S	–	S	S	–	S	–	S	S	–
Oil palm	S	S	L	S	S	S	S	L	S	S	S	S	S
Cocoa	S	–	–	S	S	S	S	–	–	S	S	S	L
Coffee	S	S	S	L	S	S	S	L	S	–	–	S	S
Paddy	S	S	S	S	S	S	S	S	S	S	S	S	S
Tobacco	–	S	S	–	–	–	–	S	L	–	L	–	S
Starfruit	S	S	L	S	L	S	S	L	S	–	S	S	S
Papaya	L	L	L	S	L	S	S	L	S	–	S	S	L
Cempedak	S	S	S	S	S	S	L	L	S	–	L	S	S
Durian	L	S	S	S	S	S	S	S	L	–	S	S	S
Sweet orange	S	S	S	S	S	S	S	S	L	–	S	S	S
Mango	S	S	S	S	L	S	S	S	S	–	L	S	S
Mangosteen	S	S	S	L	L	S	S	L	S	–	S	S	S
Pineapple	S	S	S	S	S	S	S	L	S	–	S	S	S
Jackfruit	S	S	S	S	S	S	S	S	S	–	S	S	S
Banana	S	S	L	S	L	L	S	L	S	–	S	L	S
Rambutan	S	S	S	S	S	S	S	L	S	–	S	S	S
Water melon	S	S	S	S	S	S	S	S	L	–	S	S	S

(Continued)

TABLE 4.9 (Continued)
Recommendation of Suitable Crops for Cultivation in Particular States in Malaysia

CROPS							STATES						
	Johor	Kedah	Kelantan	Melaka	Negeri Sembilan	Pahang	Perak	Perlis	Pinang	Sabah	Sarawak	Selangor	Terengganu
Chinese spinach	S	S	S	S	S	S	S	L	S	–	–	S	L
Lady's fingers	S	S	S	S	S	S	S	L	S	–	–	S	L
Chilli	S	S	S	S	S	S	S	S	S	–	–	S	S
Long bean	S	S	S	S	S	S	S	S	S	–	–	S	S
Cucumber	S	S	S	S	S	S	S	S	S	–	–	S	S
Tomato	S	L	S	S	L	S	L	L	L	–	–	S	S
Hot chilli	S	L	S	S	L	S	S	L	L	–	–	L	S
Ginger	S	L	S	L	L	S	S	L	S	–	–	S	L
Pepper	S	–	–	–	–	–	–	–	–	L	S	–	–
Lime	S	S	S	S	S	S	S	S	S	–	–	S	S
Lemon grass	S	S	S	S	S	S	S	S	S	–	–	S	S
Maize	L	S	S	L	S	S	S	S	S	–	–	S	S
Groundnut	L	L	S	L	L	S	S	L	L	–	–	L	S
Cassava	L	S	L	L	L	S	S	S	L	–	–	L	L
Sweet potato	S	L	S	S	L	L	L	L	L	–	–	S	L
Sago	L	–	L	–	–	–	–	–	–	–	S	–	–

Note: S, suitable; L, less suitable.

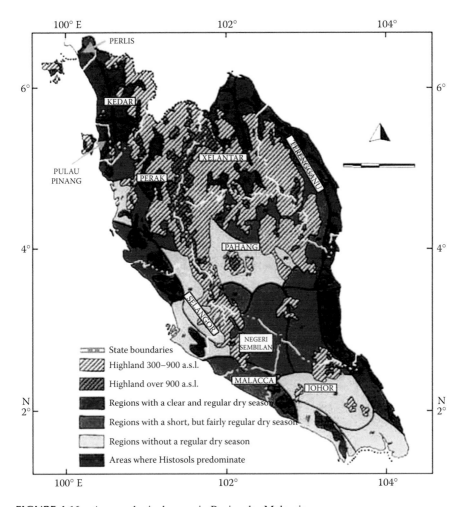

FIGURE 4.16   Agroecological zones in Peninsular Malaysia.

East Malaysia

East Malaysia, which is separated from Peninsular Malaysia by the South China Sea, consists of two large states: Sarawak and Sabah. Sarawak, which has an area of 123,466 km^2 and Sabah, which has an area of 73,711 km^2, are located at the southwest and northeast of East Malaysia, respectively. Out of the total land area, 5.31 and 2.15 million ha are arable in Sarawak and Sabah, respectively. Sarawak and Sabah have similar weather condition: moist and wet all year round especially toward the interior, with mean annual rainfall between 2500 and 5000 mm, and no months below 100 mm. AEZs characteristics, soils of Sarawak, and the agricultural suitability are summarized in Table 4.10. Sarawak's AEZs, which encompass 11 zones, are shown in Figure 4.17.

Many parts of Sabah are very steep. Among the major soils are Fluvisols and Gleysols (S1), Lithosols, Regosols and Cambisols (S3), Luvisols and Nitosols (S5),

TABLE 4.10
Characteristics of Agroecological Zones (AEZs) in Sarawak

No.	AEZ	Terrain	Slope (°)	Soil Type	Altitude (m a.s.l.)	Climate	Natural Vegetation	Irrigated	Dryland	Tree Crops
1	Coastal marshy plains	Alluvio-marine and marine plains	0–6	Thionic Histosols, Fluvisols	0–30	Moist to wet	Peat, swamp, forest	Marginal to not suitable	Marginal	Marginal
2	Inland marshy plains	Alluvio organic plains	0–6	Histosols	0–30	Moist to wet	Peat, swamp, forest, marsh	Marginal	Marginal	Marginal after reclamation
3	Saline clay plains	Alluvio-marine clay plains	0–6	Gleysols, Thionic Histosols	0–30	Moist to wet	Mangrove and swamp forest	Good–moderate	Good–marginal	Moderate
4	Lower riverine zone	Alluvio plain complex	0–6	Fluvisols, Oxisols, Podzols	0–500	Very wet, wet and moist	Riverine forest	Very good	Good–moderate (marginal to not suitable on podzols)	Good–moderate
5	Upper riverine zone	Alluvio-colluvial valley complex	0–6	Fluvisols, Gleysols, Acrisols	500–1000	Temperate, moist–wet	Riverine mountainous forest	Very good–moderate	Good–moderate	Moderate–good
6	Undulating and hilly sedimentary	Shale and sandstone hillocky and hilly slopes	6–25	Acrisols, Luvisols, Podzols	0–150	Moist–wet	Dipterocarp mixed hill forest	Moderate–unsuitable	Good–moderate	Good–moderate

(Continued)

TABLE 4.10 (*Continued*)
Characteristics of Agroecological Zones (AEZs) in Sarawak

No.	AEZ	Terrain	Slope (°)	Dominant Soil Type	Altitude (m a.s.l.)	Climate	Natural Vegetation	Irrigated	Dryland	Tree Crops
7	Igneous hilly and mountainous	Igneous hilly–mountainous uplands	6–25	Oxisols, Acrisols, Lithosols/Regosols	150–900	Very wet, moist–wet, temperate	Mixed dipterocarp hill forest	Generally unsuitable	Good–marginal	Good–moderate
8	Mountainous sedimentary	Sandstone and shale mountainous	25–33	Acrisols, Lithosols/Regosols	150–300	Wet–very wet	Mixed dipterocarp hill forest	Unsuitable	Unsuitable	Marginal–unsuitable
9	Dissected Karst	Limestone mountainous	>33	Cambisols, Lithosols/Regosols	30–300	Wet–very wet	Mixed dipterocarp hill forest	Unsuitable	Unsuitable	Marginal–unsuitable
10	Steep mountainous sedimentary	Sandstone and shale steep mountainous	>33	Acrisols, Lithosols/Regosols	300–900	Very wet, temperate	Mixed dipterocarp hill forest	Unsuitable	Unsuitable	Marginal–unsuitable
11	Steep very high mountainous	Shale and sandstone steep mountainous	>33	Lithosols/Regosols	900–1525	Very wet–moist–wet	Mixed dipterocarp hill forest	Unsuitable	Unsuitable	Generally unsuitable

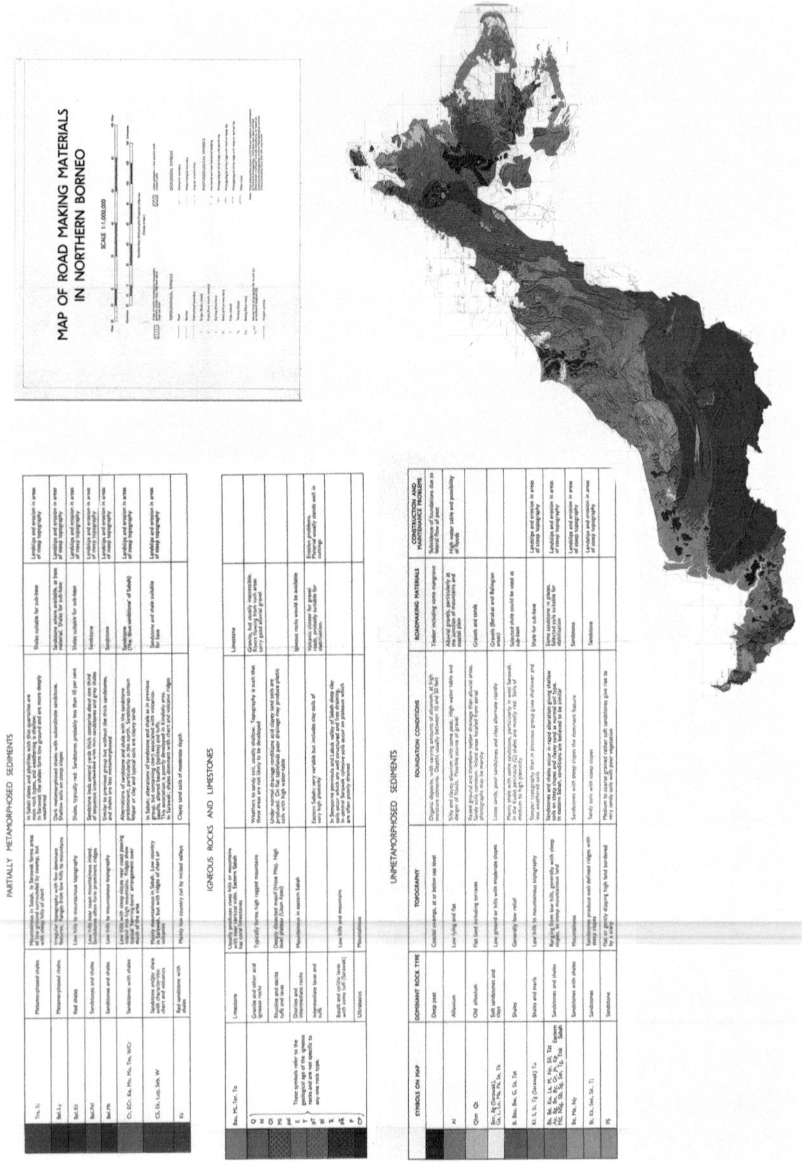

FIGURE 4.17 Agroecological zones in Sarawak.

Acrisols and Ferralsols (S6), and Histosols (S7). The codes in parenthesis are equivalent to the groups. The S1 soils group is less extensive and is widely cultivated with paddy. S3 soils group consists of all shallow soils of the highlands and the Regosols of the alluvial plains. This soils group is, hence, the most extensive. Similar to S1, the S5 soils group is also less extensive with dryland crops. The second most extensive soils group is S6 soils group, which includes all the main agricultural soils. Lastly, S7 soils group, which is mainly peat, exists in rather large blocks and is mostly used for commercial pineapple production. Huge areas of S7 soils group are also developed for oil palm and sago plantation.

SOIL MAP

Soil map provides information on distribution of land types and determines plants suitability and also the form of land management in an area. The basic unit used for mapping soil is a *soil series*. Soil classification system used in Peninsular Malaysia is based on a system of *USDA Soil Taxonomy, 1992*, which has been modified to suit local conditions. Under this classification if the characteristics of the soil do not fit the previous classified soil, the soil is given a new series name after the place in which it has been first characterized. In Peninsular Malaysia more than 240 soil series was created and reorganized into 11 groups based on the parent material, the characteristics of the main land, landscape, and way of soil formation. These groups are as follows:

1. Soil formed from igneous and metamorphic rocks of high grade
2. Soils formed from sediments and low-grade metamorphic rocks
3. Alluvial soils formed from old alluvium
4. Laterite soil
5. Alluvial soils formed from river current
6. Land formed from almost old river alluvium
7. BRIS soil
8. Potential acid sulphate soils and sulfuric acid
9. Marine clay soil without acidic sulfate
10. Organic soil (peat) shallow

Malaysia has a wide variety of soils. These soils have been mapped on mountainous, hilly, rolling, undulating, level, and swampy terrain. They occur at high and low altitudes. Both shallow moderately deep and deep soils have been recognized and mapped. Some of these soils are organic in origin, whereas most of them are made up of mineral soil materials. These soils can be well drained or poorly drained or can even be under water for long periods of time. Malaysian soils have a variety of colors ranging from blue to white, from yellow to brown, and to red. They can be sandy in texture without any clay or have a range of clay contents giving rise to sandy loam, sandy clay, loam, sandy clay, and clay textures. These soils can be developed over a range of parent materials. In Malaysia over 500 soil series have been identified. These different types of soil have developed over different topographic situations and over different parent materials or rock types and alluvial deposits. The fact that Peninsular Malaysia, Sabah, and Sarawak used different soil classification systems further complicates the identification of these soils. Sometimes what appears to be the same type of soil is given different names in the three regions (Paramananthan 2012).

REFERENCES

Ahmad, F., Yahaya, A. S., and Farooqi, M. A. (2006). Characterization and geotechnical properties of Penang residual soils with emphasis on landslides. *American Journal of Environmental Sciences*, 2(4), 121.

American Society for Testing and Materials (2013). D4427-13 Standard Classification of Peat Samples by Laboratory Testing.

Anuar, K. and Ali, F. H. (1997). Reinforced modular bock wall with residual soils as backfill material. *Proceedings 4th Regional Conference in Geotechnical Engineering (GEOTROPIKA'97)*, Johor Bahru, Malaysia, 425–431.

Das, B. M. (2004). *Principles of Foundation Engineering*. 5th edition. Pacific Grove, CA: Thomson Brooks/Cole Publishing Company.

Duraisamy, Y., Huat, B. B., and Aziz, A. A. (2007). Engineering properties and compressibility behavior of tropical peat soil. *American Journal of Applied Sciences*, 4(10), 768.

Fao, F. (1988). UNESCO soil map of the world, revised legend. World Resources Report, 60, 138.

Fong, S. S. and Mohamed, M. (2007). Chemical characterization of humic substances occurring in the peats of Sarawak, Malaysia. *Organic Geochemistry*, 38(6), 967–976.

Huat, B. (2006). Deformation and shear strength characteristics of some tropical peat and organic soils. *Pertanika Journal of Science & Technology*, 14(1–2), 61–74.

Huat, B. B., Prasad, A., Asadi, A., and Kazemian, S. (2014). *Geotechnics of Organic Soils and Peat*. Boca Raton, FL: CRC Press.

Huat, B. H., Sina, K., Arun, P., and Maassoumeh, B. (2011). State of an art review of peat: General perspective. *International Journal of Physical Sciences*, 6(8), 1988–1996.

Indraratna, B., Balasubramaniam, A. S., and Balachandran, S. (1992). Performance of test embankment constructed to failure on soft marine clay. *Journal of Geotechnical Engineering*, 118(1), 12–1.

Kepli, M. I. (1994). Properties of Granite Derived Residual Soils, UITM Project Report.

Kolay, P. K., Sii, H. Y., and Taib, S. N. L. (2011). Tropical peat soil stabilization using class F pond ash from coal fired power plant. *International Journal of Civil and Environmental Engineering*, 3(2), 79–83.

Little, A. L. (1969). The engineering classification of residual tropical soils. *Proceeding of Soil Mechanics & Foundation Engineering Conference*, Mexico.

Maail, S., Huat, B. B., Jamaludin, S., Sew, G. S., and Ali, F. H. (2004). Index, engineering properties and classification of tropical residual soils. *Tropical Residual Soils Engineering*, 37–55.

Marto, A. and Kasim, F. (2003). Characterisation of Malaysian residual soils for geotechnical and construction engineering. Research Report, Research Management Centre, UTM.

Marto, A., Mohd. Yunus, N. Z., Pakir, F., Latifi, N., Nor, A. H. M., and Tan, C. S. (2015). Stabilization of marine clay by biomass silica (non-traditional) stabilizers. *Applied Mechanics and Materials*, 695, 93–97.

Melling, L., Hatano, R., and Goh, K. J. (2005). Soil CO_2 flux from three ecosystems in tropical peatland of Sarawak, Malaysia. *Tellus B*, 57(1), 1–11.

Mohamad, E. T., Alshameri, B. A., Kassim, K. A., and Gofar, N. (2011). Shear strength behaviour for older alluvium under different moisture content. *Electronic Journal of Geotechnical Engineering*, 16, 605–617.

Mohamad, H., Choon, T., Azam, T., and Tong, S. (1995). The Petronas Towers-The Tallest Building in the World. In Habitat and the High-Rise. Tradition and Innovation, Fifth World Congress (pp. 14–19).

Pansu, M. and Gautheyrou, J. (2007). Handbook of Soil Analysis: Mineralogical, Organic and Inorganic Methods. Berlin, Germany: Springer.

Paramananthan, S. (2012). Keys to the Identification of Malaysian Soils using Parent Materials, Param Agricultural Soil Surveys (M) Sdn. *Bhd., Malaysia*, 2–20.

Paramananthan, S. (2000). *Soils of Malaysia: Their Characteristics and Identification*, Vol. 1. Kuala Lumpur, Malaysia: Academy of Sciences Malaysia.

Paramananthan, S. (1978). Register of Soils-Peninsular Malaysia. Kuala Lumpur, Malaysia: Soils and Analytical Services Bulletin-Kementerian Pertanian Malaysia.

Pushparajah, E. and Amin, L. L. (1977). *Soils under Hevea in Peninsular Malaysia and their Management*, Kuala Lumpur, Malaysia: Rubber Research Institute of Malaysia.

Salih, A. G. et al. (2012). Review on granitic residual soils. *Electronic Journal of Geotechnical Engineering*. 16(T), 2645–2658.

Singh, H. and Huat, B. B. K. (2004). Terra firma and Foundations. In *4th International Conference on Landslides, Slope Stability and Safety of Infrastructure*. March 24–25, Kuala Lumpur, Malaysia.

Staub, J. R. and Gastaldo, R. A. (2003). Late Quaternary sedimentation and peat development in the Rajang River Delta, Sarawak, East Malaysia. Tropical Deltas of Southeast Asia Sedimentology, Stratigraphy, and Petroleum Geology SEPM Special Publication No. 76, 71–87.

Suratman, S. (2004). IWRM: Managing the groundwater component. In *Malaysia Water Forum*, Kuala Lumpur, Malaysia.

Soil Survey Staff (1994). Keys to soil taxonomy. Soil Conservation Service. U.S. Department of Agriculture and Natural Resource Conservation Service, U.S. Government Printing Office, Washington, DC.

Taha, M. R., Desa, H. M., and Kabir, H. (2002). The use of granite residual soils as a landfill liner material. *Proceedings 4th Regional Conference in Geotechnical Engineering (GEOTROPIKA'97)*, Johor Bahru, Malaysia, 264–267.

Tan, B. K., Huat, B. B. K., Sew, G. S., and Ali, F. H. (2004). Country case study: Engineering geology of tropical residual soils in Malaysia. *Tropical Residual Soils Engineering*, 237–244.

Teng, C. S. (2004). Keys to soil classification in Sarawak. Department of Agriculture, Sarawak, Malaysia.

Wesley, L. D. (2010). *Geotechnical Engineering in Residual Soils*. Hoboken, NJ: John Wiley & Sons.

Wetlands International (2010). A quick scan of peatlands in Malaysia. Wetlands International-Malaysia: Petaling Jaya, Malaysia. 50.

Wüst, Raphael, A.J., and Bustin, R. Marc. (2004). Late Pleistocene and Holocene development of the interior peat-accumulating basin of tropical Tasik Bera, Peninsular Malaysia. *Palaeogeography, Palaeoclimatology, Palaeoecology*, 211(3–4), 241–270.

Yulindasari, I. (2006). Compressibility characteristics of fibrous peat soil. Master Dissertation, Universiti Teknologi Malaysia, Malaysia.

Zainorabidin, A. and Wijeyesekera, D. C. (2007). Geotechnical Challenges with Malaysian Peat. Advances in Computing and Technology, *Proceedings of AC&T 2007*. London: ICGeS Press, 2007.

5 Soil Properties (Physical, Chemical, Biological, Mechanical)

*Christopher Teh Boon Sung, Che Fauziah Ishak,
Rosazlin Abdullah, Radziah Othman, Qurban
Ali Panhwar, and Md. Maniruzzaman A. Aziz*

CONTENTS

INTRODUCTION

Soil is an anchor for plant roots and as a water holding reservoir for needed moisture, soil delivers a hospitable place for a plant to take root. Some of the soil properties affecting plant growth include soil texture (coarse or fine), aggregate size, porosity, aeration (permeability), and water holding capacity. A significant function of soil is to store and supply nutrients to plants. The capability to perform this function is referred to as soil fertility. The clay and organic matter content of a soil directly influence its fertility. Greater clay and organic matter content will generally lead to greater soil fertility. The rate of water movement into the soil (infiltration) is influenced by its texture, physical condition (soil structure and tilth), and the amount of vegetative cover on the soil surface. Coarse (sandy) soils allow rapid infiltration but have less water storage ability, due to their usually large pore sizes. Fine-textured soils have an abundance of micropores, allowing them to retain a lot of water but also causing a slow rate of water infiltration. Organic matter tends to increase the ability of all soils to retain water and also upsurges infiltration rates of fine-textured soils.

Soil resources serve as a basis for food security. Soil properties, together with climate, govern what type of plants will grow in a soil or what particular crops will grow in a region. The properties of a soil play a big part in determining the plant's ability to extract water and nutrients. If plants are to grow to their maximum yield potential, the soil must provide a conducive or satisfactory condition for plants to grow.

Malaysia has an extensive variety of soils. These soils have been mapped on mountainous, hilly, rolling, undulating, level, and swampy terrain. So far over 500 soil series have been identified in Malaysia. These various soil types have developed over different topographic conditions and over different parent materials or rock types and alluvial deposits. The fact that Peninsular Malaysia, Sabah, and Sarawak use different soil classification systems further complicates the identification of these soils (Paramananthan, 2012). Residual soils are developed from the weathering process of rocks. Although several researches have been shown on engineering properties of residual soils, the study on mineralogy, microstructure, microfabric, and chemical composition of residual soils, despite of its status, is still lacking. In addition, correlations developed between the engineering properties of granitic residual soils are also still lacking. The generated correlations can be used as guides for preliminary designs for geotechnical structures created on or in residual soil of Peninsular Malaysia (Amination and Fauziah, 2003).

PHYSICAL PROPERTIES OF MALAYSIAN SOILS

SOIL PHYSICAL PROPERTIES

Soil physical properties refer to properties such as soil texture, bulk density, aggregation, aggregate stability, and soil water content and water retention. Malaysian soils vary widely in texture from as low as 3% (sandy soils) to over 90% (clayey soils) of clay content. The mean sand content of Malaysian soils is 41%, which is nearly the same as that of the clay content, 43%. Bulk density is an indication of soil compaction, which is determined as the weight of dry soil per unit soil volume. Soil compaction is highly dependent on soil management practices, but typically, Malaysian soils have bulk density values ranging from 0.8 to 1.9 Mg/m^3, although peat soils have much lower bulk density values, as low as 0.09 Mg/m^3, depending on the organic matter types and their degree of decomposition.

Aggregation refers to the distribution of aggregate sizes, whereas aggregate stability is the resistance of these aggregates to withstand the disruptive forces from water or wind. Aggregation and aggregate stability are important soil physical properties because they indicate not only soil fertility but also how well the soil can resist erosion. In agriculture, we desire aggregates that are not unstable as they will easily crumble and aggregates that are not too stable as they behave like stones or rocks, which can complicate field-planting practices.

Organic matter is one key ingredient in the soil that affects aggregation and aggregate stability. More specifically, the organic matter components, primarily fulvic acids and humic acids, have differing impacts on aggregate stability. The effects of organic matter constituents vary between temperate and tropical soils. In tropical soils, fulvic acids are more effective in increasing aggregate stability than humic acids, most probably because there are more fulvic acids in tropical soils than in temperate soils (which have more humic acids than in tropical soils). In Peninsular Malaysia, for example, 75%–90% of the organic carbon are fulvic acids (Zainab, 1977). Because of the higher organic matter turnover rate in tropical soils than in temperate soils (Greenland et al., 1992), the humic acids are converted into fulvic acids at a faster rate in tropical soils. Other important factors are free iron and aluminum oxides and exchangeable cations.

Teh (2012a) used multiple linear regression to show that silt, followed by free Fe oxides, fine sand, fulvic acids, then humic acids were the most important soil constituents to explain the observed differences in aggregate stability between four Malaysian soil types (Ultisols and Oxisols). Moreover, the physical and chemical properties of individual aggregate size fractions are often different from one other. The amount of clay, organic matter, and cations, for example, often differs from one aggregate size fraction to another. Teh (2012b) and Teh et al. (2005) observed that as the aggregate size decreased, the amount of clay, silt, organic matter, and free Fe oxides would increase, and the aggregation and the amount of sand would decrease. Generally, it was observed that aggregate stability would increase with decreasing aggregate size until aggregate size fraction becomes <0.3 mm below which the stabilities of aggregate size fractions between soils were generally similar to one another. Table 5.1 shows the mean soil physical properties of some Malaysian soils.

Another important soil physical property is the soil water content and the water retention. There are three important points in the soil water content: saturation, field capacity, and permanent wilting point. Saturation is the point of maximum amount of water that a soil can hold below which gravity will be able to pull away the water from the soil until a certain point known as the field capacity. At this point, water is more tightly bound to the soil and held stronger by the soil than the pull of gravity. At the permanent wilting point, water is held too strongly by the soil that even plant roots are unable to obtain the water. Consequently, the amount of soil water between field capacity and permanent wilting point is known as the available water content. Different soils have different saturation, field capacity, and permanent wilting points, depending on the soil compaction, organic matter content, and soil texture. For Malaysian soils, the volumetric soil water content at saturation, field capacity, and permanent wilting point can range between 36%–89%, 10%–67%, and 3%–49%, respectively, giving the available volumetric soil water content between 1% and 13%. Generally, the soil water content at permanent wilting point is more influenced by soil texture than soil management practices, whereas the latter plays a greater role in affecting the soil water content at saturation and field capacity.

Instead of measuring the soil water retention, we can estimate it by using the Saxton and Rawls (2006) set of equations. However, Teh and Iba (2010) calibrated

TABLE 5.1

Mean Physical Soil Properties (and Organic Carbon Content) for Some Malaysian Soils

Series	%Sand 0.05–2 mm	%Clay <2 μm	%OC	%Volumetric SAT	FC	PWP	mm/hour Ks[a]
Awang	75.1	14.6	0.3	47.9	19.4	11.4	70.9
Batu Anam	19.8	35.1	0.2	55.3	42.2	24.5	5.9
Batu Lapan	31.1	56.7	1.5	60.9	42.7	35.8	24.9
Briah	16.4	54.4	1.0	70.8	57.9	38.4	7.0
Bukit Tuku	34.9	26.0	0.3	50.1	27.5	20.8	27.6
Bungor	49.5	40.2	0.5	53.9	36.8	30.0	18.6
Chat	12.0	64.6	1.0	58.4	47.3	39.2	3.2
Cherang Hangus	62.2	27.7	0.4	46.7	28.7	21.3	16.0
Chuping	35.4	33.3	0.3	76.1	35.7	20.8	157.6
Durian	30.6	49.3	0.8	55.4	42.6	34.7	5.1
Gong Chenak	38.5	41.7	0.4	49.7	37.2	29.4	5.6
Halu	53.2	25.2	0.3	60.0	44.6	37.0	17.3
Harimau	61.8	34.7	0.8	54.7	27.3	20.8	67.0
Holyrood	61.8	29.0	0.7	76.5	28.8	16.2	244.6
Jintan	18.7	60.8	0.2	69.4	47.3	37.1	23.7
Kampong Pusu	43.3	34.3	1.8	54.5	34.4	28.5	27.0
Kaning	47.7	43.7	0.3	58.2	27.2	24.4	61.6
Katong	5.1	87.5	0.9	62.3	51.5	46.8	3.9
Kawang	56.1	28.5	0.6	43.8	30.0	22.0	6.1
Kuantan	13.2	65.0	0.5	71.7	35.4	27.9	98.3
Kulai	30.1	37.2	0.4	58.9	39.4	32.0	18.1
Lambak	36.7	59.2	0.4	46.6	25.3	21.3	20.4
Lanchang	25.9	59.9	0.3	71.0	47.0	43.1	27.8
Langkawi	23.9	69.1	0.4	77.2	43.0	32.8	83.9
Lintang	77.8	17.5	0.3	78.2	20.1	12.7	405.6
Lunas	56.1	36.2	0.2	50.3	35.6	26.6	9.2
Marang	61.0	25.8	0.2	48.1	29.9	23.9	20.1
Munchong	10.6	80.7	0.9	63.1	50.2	44.7	5.5
Nangka	71.4	14.9	0.7	53.6	22.4	13.1	88.2
Napai	68.2	12.2	1.7	59.9	20.3	10.7	140.0
Padang Besar	54.3	39.0	1.2	50.2	34.5	29.1	9.2
Patang	32.7	39.0	0.6	58.9	38.8	30.5	18.3
Penambang	56.9	20.9	1.3	49.5	36.2	30.4	5.2
Prang	22.7	65.6	0.9	64.7	44.3	35.2	28.9
Rasau	35.4	29.7	1.2	69.4	34.2	20.0	113.6
Rengam	44.9	49.3	1.5	55.2	38.3	31.6	13.4
Ringlet	42.1	36.1	0.5	56.3	36.8	31.2	16.4
Sagu	19.5	64.0	0.1	72.7	43.1	39.6	52.5
Segamat	6.4	72.1	0.9	69.6	44.6	39.1	40.9

(Continued)

TABLE 5.1 (*Continued*)
Mean Physical Soil Properties (and Organic Carbon Content) for Some Malaysian Soils

	%Sand	%Clay		%Volumetric			mm/hour
Series	**0.05–2 mm**	**<2 µm**	**%OC**	**SAT**	**FC**	**PWP**	**Ks[a]**
Senai	23.0	56.9	0.8	56.2	37.6	32.2	14.4
Serdang	68.2	28.2	0.6	52.8	29.7	21.3	35.7
Sg. Buloh	92.5	3.1	0.5	41.7	8.1	4.7	86.4
Sg. Mas	41.3	37.1	0.5	57.0	39.2	31.3	15.6
Sogomana	16.9	52.6	0.6	56.7	41.6	31.3	9.1
Subang	72.9	12.1	1.2	54.1	23.2	12.5	72.3
Tampin	51.3	32.3	0.4	46.2	28.4	18.5	14.5
Tanah Rata	64.3	19.3	0.6	63.8	47.7	35.2	10.2
Tavy	26.0	60.7	0.9	51.5	29.9	23.0	35.2
Tebok	36.6	38.1	0.6	59.5	33.2	24.3	75.1
Ulu Dong	34.8	56.8	0.3	65.7	36.9	34.6	64.1
Ulu Tiram	65.7	30.6	0.6	47.1	29.3	22.4	12.7

Source: Maene, L. et al., *Register of Soil Physical Properties of Malaysian Soils.* Technical Bulletin, Faculty of Agriculture, Universiti Pertanian Malaysia, Serdang, Malaysia, 1983.
OC–organic carbon; SAT, FC, and PWP–the soil water content at saturation, field capacity, and permanent wilting point, respectively; and Ks–saturated hydraulic conductivity.
[a] Ks is estimated (From Saxton, K.E. and Rawls, W.J., *Soil Sci. Soc. Am. J.*, 70, 1569–1578, 2006).

the Saxton and Rawls (2006) equations for Malaysian soils. Their developed equation for calibration was as follows:

$$\hat{P}_i = a \cdot P_i (1 - P_i)$$

where P_i and $\hat{P}_i$ are the uncalibrated and calibrated estimated values, respectively, for soil sample no. i, and the parameter a values were 2.225, 1.605, and 1.528 for saturation, field capacity, and permanent wilting point, respectively. The calibrated method was validated against three independent soil datasets. Validation tests showed that the calibrated method remained stable and was more accurate than that without calibration by an average of between 8% and 49%.

The soil water retention also provides us information on the point of soil water content below which the crop begins to feel the effects of water stress as presented in Figure 5.1.

The relative soil water content (RWC; m³/m³) is calculated by

$$RWC = \frac{\Theta_v - \Theta_{v,pwp}}{\Theta_{v,sat} - \Theta_{v,pwp}}$$

where:

Θ_v is the volumetric soil water content (m³/m³)

$\Theta_{v,sat}$ and $\Theta_{v,pwp}$ are the volumetric soil water contents at saturation and permanent wilting point, respectively (m³/m³)

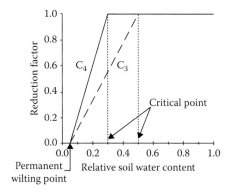

FIGURE 5.1   Plants will start to be water stressed at soil water content below the critical point. C_4 plants are less sensitive than C_3 plants to water stress.

The soil's critical point $\Theta_{v,cr}$ (m³/m³) below which the plant begins to encounter the effects of water stress is some point along the relative soil water content, and it can be determined by

$$\Theta_{v,cr} = \Theta_{v,pwp} + p\left(\Theta_{v,sat} - \Theta_{v,pwp}\right)$$

where p is 0.5 and 0.3 for C_3 and C_4 plants, respectively.

Soil physical properties often respond slower than soil chemical properties due to organic matter amendments. For instance, no significant changes were observed for soil bulk density and total porosity even after 3–10 years of soil organic mulching, as reported by Acosta-Martinez et al. (1999), Bescansa et al. (2006), Karlen et al. (1994), and Onweremadu et al. (2007). Similarly, Teh and Zauyah (2001) observed no difference in soil physical changes (such as bulk density, aggregation, and aggregate stability) even after 10 years of annual mulch application using empty fruit bunches (EFB).

Nonetheless, Moraidi et al. (2015) reported that annual EFB application as a mulching material for 3 years improved nearly all soil physical and chemical properties that were measured. They compared EFB with oil palm fronds and an EFB mat (called as Ecomat). EFB was found to be the most effective material to increase soil aggregation, aggregate stability, soil water retention at field capacity, available soil water content, and the relative proportion of soil mesopores. Due to these improved soil physical properties, EFB also gave the highest soil water content. Unlike Ecomat and oil palm fronds that concentrated more water in the upper soil layers, EFB distributed the soil water more uniformly throughout the whole soil profile.

Malaysia experiences high rainfall, where the country's annual rainfall is 2000–3000 mm. Consequently, water erosion is of particular importance, whereas erosion by wind, in contrast, is of little importance because the soil tends to be continuously wet from frequent and heavy rainfall, and Malaysia experiences slow average daily wind speeds, typically only about 2 m/s.

Improving the soil physical properties is one of the most important methods to reduce soil erosion. Soil erodibility is often represented as the K factor in the RUSLE soil erosion model, where smaller K values denote that a soil has a stronger resistance

TABLE 5.2
Range of Soil Erodibility Based on Soil Texture of Peninsular Malaysia Soil Series

Soil Texture	K Factor (tonne ha)(ha hour/MJ mm)
Clay	0.042–0.065
Clay loam	0.030–0.047
Sandy clay	0.031–0.043
Sandy clay loam	0.028–0.059
Sandy loam	0.004–0.036
Silt loam	0.014–0.027
Silty clay loam	0.032

Source: Yusof, M.F. et al., Modified soil erodibility factor, K, for Peninsular Malaysia soil series, in *3rd International Conference on Managing Rivers in the 21st Century: Sustainable Solutions for Global Crisi of Flooding*, Pollution and Water Scarcity (RIVER 2011), Penang, Malaysia, December 6–9, 2011, pp. 799–808, 2011.

against erosion, whereas larger K values denote a weaker soil against erosion. K factor depends on soil texture, organic carbon content, soil structure, and soil permeability. Using the soil properties of 76 soil series in Peninsular Malaysia, Yusof et al. (2011) developed a range of K values for Malaysian soils based only on their soil texture as presented in Table 5.2. This table shows that soils with higher clay content are more susceptible to erosion than soils having higher silt and sand content. For instance, Yusof et al. (2011) determined that 56% of the 74 soil series they used had high clay content such as Akob, Batu Anam, and Chengai, which were more erodible than soils with higher silt or sand content (e.g., Rudua, Holyrood, Lunas, and Marang).

In Malaysia, soil erosion is controlled using three basic principles: (1) agronomic methods, (2) soil management methods, and (3) mechanical methods. Agronomic methods include the use of vegetation to protect the soil, and soil management methods are ways to improve soil fertility and soil structure to increase soil resistance against erosion, and mechanical methods are the use of physical and artificial structures such as walls and hill terraces to reduce soil erosion.

Agronomic and soil management methods are cheaper, but they require time before they can give full or effective soil protection. The key to agronomic methods is to provide physical cover to the ground, covering it against rain splash erosion and soil and water loss via runoff. It is not necessary to have full ground cover to have a significant reduction in soil losses: 70% ground cover is sufficient (Morgan, 2005). In Malaysia, legumes are typically used as cover crops. Popular cover crops are *Pueraria phaseoloides*, *Calopogonium mucunoides*, *Centrosema pubescens*, and *Mucuna bracteata*. These legumes are popular because they are fast growing, able to fix N, and provide thick and complete ground cover. Vetiver grass is also a popular grass typically planted along hill slopes in urban areas. This grass has a very deep, wide, and dense rooting system (>2 m deep).

TABLE 5.3
Isohumic Factor of Several Types of Organic Materials

Organic Material	Isohumic Factor
Plant foliage	0.20
Green manures	0.25
Cereal straw	0.30
Roots of crops	0.35
Farmyard manure	0.50
Deciduous tree litter	0.60
Coniferous tree litter	0.65
Peat moss	0.85

Source: Kolenbrander, G.J., *Trans. Int. Congr. Soil Sci.*, 2, 129–136, 1974.

The aim of soil management is to improve or maintain high soil fertility and soil structure levels so that the soil is able to support a good growth of plants, thus reducing the impact of rain drops, runoff, and wind. Addition of organic matter is vital because its addition will increase, among others, the water retention capacity and higher aggregate stability. Green manures decompose rapidly, but its rapid beneficial effects on the soil are short lived. Straw decomposes less rapidly, so its beneficial effects are less rapid but more prolonged. The slowest to decompose is manure. A measure of the quantity of humus produced per unit quantity of organic matter is called isohumic factor as given in Table 5.3.

Moraidi et al. (2013) reported that after four consecutive annual applications of EFB mulch, 23.42 kg C was added to the unit area of the land surface of which only 12% is converted into the soil organic C at 0–0.30 m depth. This corresponds to 1.0 million tonne/year EFB C sequestrated into the soil globally, of which Malaysia contributes 35%.

Mechanical methods include structures such as walls, gabions, and terraces. Terraces are built on hill slopes to reduce the length and slope of hills, thus reducing the amount and speed of runoff. However, terraces involve the use of heavy machinery that can compact the soil and involve the removal of the fertile topsoil. Consequently, some plantations in Malaysia have put an end to the use of terraces and built silt pits instead. These so-called silt pits are long, narrow, and close-ended soil trenches, built in perpendicular to the hill slope direction. The idea is for the silt pits to capture the runoff during the period of rainfall, thus trapping the water, sediments, and nutrients from the runoff that would otherwise be lost from the hill.

Bohluli et al. (2014) was one of the few studies that examined not only the effectiveness of silt pit as a soil and water conservation method but also the effect of silt pit size. They examined four dimensions (width × length × depth in meters) of silt pit: $1 \times 3 \times 1$, $1.5 \times 3 \times 1$, $2 \times 3 \times 1$, and $2 \times 3 \times 0.5$. They also compared the effects of these silt pit sizes on the soil with control (no silt pit and with oil palm fronds as mulch). They determined that silt pits increased soil water content between 3% and 19% compared to control and that the silt pits also conserved more soil nutrients than the control. Results showed that the silt pit with the smallest opening area conserved more soil water content in oil palm active root zone and is the best effect to improve soil chemical parameters inside and outside of the pit compared with other treatments. This was because pits with smaller opening area had bigger wall-to-floor area (W:F) ratio, which caused higher lateral water infiltration through silt pit's walls than water percolation through silt pit's floor area. Moreover, silt pits with narrower opening area helped the water head to be higher than other wider pits and redistributed more dissolved nutrients into the topsoil.

CHEMICAL PROPERTIES OF MALAYSIAN SOIL

The two main soil chemical properties controlling chemical processes in the soil system are cation exchange capacity (CEC) and soil acidity.

CATION EXCHANGE CAPACITY

In the soil system, the main soil component controlling chemical reactivity is clay, due to its small particulate size (<2 µm) and high surface area. The adsorption process, which is the main chemical process controlling the availability of nutrients to the plant system takes place at the clay colloidal surface. Isomorphous substitution or the substitution by another cation of similar size and form but of lower oxidation state, during the formation of silicate clays, resulted in negative charges on the clay surfaces. The adsorption of cations on the clay surfaces is necessary to balance these electrostatic charges. The other soil component responsible for cation retention in the soil system is the organic colloids. The retention of the cations on the clay surfaces or organic colloids is termed as the CEC.

The CEC is highly related to the mineralogical composition of the soil. The CEC of the soil is influenced more by the type of mineral rather than the amount of mineral present in the soil system. Table 5.4 presents the common clay types.

TABLE 5.4
Examples of Some Common Different Clay Types

No.	Clay	CEC me /100 g
1	Organic matter	150–500
2	Kaolinite	3–15
3	Aluminum and ferum hydroxide	4
4	Illite	10–40

SOIL ACIDITY

Besides CEC, the other main soil chemical process controlling reaction in the soil system of highly weathered tropical soils is soil acidity. Soil acidification is a natural process and has long been known as a form of chemical degradation of the soil. Soil acidity can either be accelerated by the activity of plants, animals, and humans, or it can be impeded by sound management practices. Brady and Weil (1999) described soil acidity in the following manner.

> No other single soil characteristic is more important in determining the chemical environment of higher plants and soil microbes than the pH. There are few reactions involving any component of the soil or of its biological inhabitants that are not sensitive to soil pH. The sensitivity must be recognized in any soil management system.

PROCESSES OF ACID GENERATION IN SOILS

Processes of acid generation in soils can be broadly grouped into two categories: (1) those occurring under natural ecosystems through weathering processes, pyrite oxidation in acid sulfate soil, and organic matter deposition or accumulation forming peat and (2) those occurring under managed ecosystems through farming activities. The most significant H^+ and hydroxyl ion (OH^-) generating processes occur during the biogeochemical cycling of C, N, and S. Although these processes occur both under natural and managed ecosystems, under the latter system, these processes are accelerated by the activities of humans through intensive land-based crop and animal production. Overall, the main contribution toward soil acidification is from the utilization of the ammoniacal forms of N in fertilizers and legumes, which are oxidized to nitric and other acids, and in crop removal of basic cations (nutrients mining) and acid rain. In rubber plantations of Malaysia, it has been reported that due to more than 100 years of ammonium sulfate application, the soil pH had been reduced by 1 unit to about pH of 4.00 (Shamsuddin et al., 2015).

NATURAL ECOSYSTEMS

HIGHLY WEATHERED SOILS

Malaysian soils dominantly (about 70%) fall into the Ultisol and Oxisol Orders in Soil Taxonomy. These soils are acidic in nature, with pH values ranging from 4 to 5. These soils contain mainly sesquioxides and kaolinite both of which are essentially variable charge minerals. The phosphorus present in the soil system is highly fixed by the sesquioxides in the soil system, and thus these soils are extremely poor in available phosphorus. In fact, phosphorus is considered as the most limiting nutrient for crop production in the tropics. These soils also have very low basic cation status and effective CEC as presented in Table 5.5. Accessions of acidity by these soils contribute to soil degradation as a result of reactions, which liberate toxic levels of Al^{3+} and Mn^{2+}, reduce the CEC, increase the anion exchange capacity, and promote

TABLE 5.5

Selected Chemical Properties of Ultisols and Oxisols

Series	pH H_2O	pH $CaCl_2$	Exch. Cations (cmol$_c$/kg) Na	K	Ca	Mg	% bas. sat.	% Al sat.	CEC (cmol$_c$/kg)	Exch. Al (cmol$_c$/kg)	O.M (%)
Ultisols											
Rengam	4.3	3.8	0.09	0.08	0.23	0.11	10.41	26.65	4.90	1.09	1.98
Serdang	4.6	3.8	0.06	0.05	0.02	0.06	5.85	26.46	3.25	0.86	1.05
Bungor	4.1	3.6	0.07	0.15	0.24	0.13	18.15	52.33	6.00	3.14	1.45
Kuala Brang	4.4	3.9	0.06	0.17	0.18	0.07	5.05	55.89	9.50	5.31	2.03
Lanchang	4.5	4.1	0.06	0.15	0.12	0.18	5.71	15.23	8.93	1.36	2.78
Oxisols											
Segamat	4.9	4.6	0.04	0.12	1.10	0.96	24.18	3.38	9.18	0.31	3.00
Sg. Mas	4.6	3.9	0.04	0.18	0.24	0.46	12.11	8.64	7.18	0.62	1.60
Muchong	4.4	4.1	0.06	0.30	0.84	0.14	14.81	16.68	9.05	1.51	2.41

Source: Shamshuddin, J., *Pertanika,* 12, 109–111, 1989.

the loss of basic cations (Ca^{2+} and Mg^{2+}) by leaching. The activities of soil organisms are generally reduced under such conditions. This can take a toll on crop yield and impair biological N fixation. Therefore, amelioration involves both the neutralization of exchangeable Al and Mn and restoration of higher levels of exchangeable basic cations, namely Ca, throughout the soil profile.

ACID SULFATE SOILS

In Malaysia, the area covered by these soils is quite extensive about 0.5 million ha most of which is still in its pristine condition under waterlogged environments as shown in map in Figure 5.2. The pyritization process that began in the past is still ongoing in the coastal plains of Peninsular Malaysia. Pyrite forms when sulfate from seawater and ferric ions from marine sediments are reduced to sulfide and ferrous ions, respectively. These reactions occur under extremely reducing or anaerobic condition where microorganisms feeding on the organic matter present in the sediment play an important role in the reduction process.

Under flooded conditions and in the presence of organic matter, ferric ions in the seawater are readily reduced to ferrous ions with the help of microbes. The organic matter needed for the reduction process will be provided by the native vegetation. The chemical characteristics of selected acid sulfate soils are shown in Table 5.6.

Acid sulfate soils are highly buffered at pH 5–9. At this pH range, Al plays an important role in the buffering process. Also, the high Fe content in some acid sulfate soils can also buffer the acid sulfate soils from pH changes. Thus, acid sulfate soils require high amount of lime to increase the pH by just one unit.

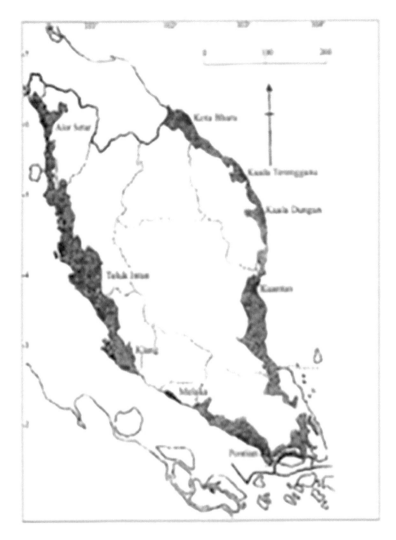

FIGURE 5.2   A map of Peninsular Malaysia showing the distribution of acid sulfate soils. (Courtesy of the Department of Agriculture, Peninsular Malaysia.)

PEATLAND

In Malaysia, peatland occupies 2.7 million ha, accounting for about 8% of the total land area of the country. The state of Sarawak in East Malaysia has the largest area of peat in the country, covering about 1.66 million ha (65% of the total peat area). Tropical peatlands make a significant contribution to terrestrial C storage because of their considerable thickness, high C content, and most importantly, their rapid peat and C accumulation rates that have often exceeded those of boreal and temperate peatlands (Figure 5.3).

TABLE 5.6

Selected Chemical Properties of Typical Sulfate Soils of Peninsular Malaysia

Series	Depth	pH	EC (mS/ cm)	Exchange Cations (cmol$_c$/kg) Na	K	Ca	Mg	% bas. sat.	% Al sat.	CEC cmol$_c$/kg	Exchange Al cmol$_c$/kg	Water-Soluble SO$_4^{2-}$-%
Kranji	0–22	6.6	8.00	41.8	4.2	9.1	26.1	226	0	35.8	0	0.48
	>63	7.1	12.00	58.3	5.1	19.1	26.4	379	0	28.7	0	0.96
Telok	0–17	4.1	0.11	0.2	0.3	2.7	4.3	24.5	26.14	30.6	8.0	0.03
	110–130	2.2	2.50	0.8	0.4	1.8	6.2	32.25	17.41	24.7	4.3	0.58
Jawa	0–5	4.2	0.14	0.7	0.7	4.2	6.0	52.48	11.31	22.1	2.5	0.01
	130–150	2.4	4.01	7.4	0.3	6.0	14.0	92.64	56.86	29.9	17.0	10.42
Sedu	0–5	4.0	0.25	0.6	0.8	1.9	3.8	32.42	16.89	21.9	3.7	0.01
	125–162	2.4	3.59	5.0	0.6	4.7	14.7	92.59	58.15	27.0	15.7	8.78

Source: Shamshuddin, J. (ed.), *Acid Sulphate Soils in Malaysia*, UPM Press, Serdang, Malaysia, 2006.

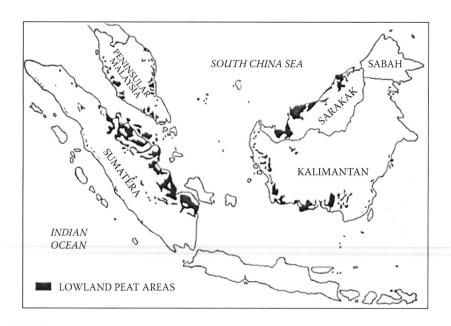

FIGURE 5.3 Distribution of lowland peatlands in Malaysia and Indonesia.

Tropical peat soils are generally very acidic with pH values of less than 4.0. The peat soil is highly buffered with respect to pH changes. The buffering capacity of the peat soil is primarily due to the carboxyl and phenolic functional groups in the humic substances of organic matter as given in Table 5.7.

TABLE 5.7

Selected Chemical Properties of Peat and Their Liming Requirement

Peat	Depth (cm)	pH	% HA	Buffer Capacity[a] mole (OH)/kg	LR[b] (tonne CaCO$_3$/ha)
1	0–15	4.53	21.23	1.95	2.74
2	0–15	3.41	26.88	1.91	18.59
3	0–15	3.93	18.30	0.71	12.78
4	15–30	3.51	25.43	0.97	18.07
5	0–15	3.42	23.78	0.91	15.42
6	15–30	3.86	8.40	0.02	10.14
7	0–15	3.37	35.67	2.26	22.29
8	0–15	3.48	20.68	0.10	13.31
9	0–15	3.36	23.45	1.09	15.42
10	0–20	3.38	39.52	5.94	25.99
11	20–40	3.17	12.38	0.48	16.48
12	0–15	3.34	19.82	0.34	15.94
13	20–40	4.54	8.30	0.04	1.68
14	0–15	3.43	26.92	4.56	21.24

[a] Buffer capacity ($\times 1000$).

[b] Lime requirement to attain a pH of 5.0.

WAYS TO OVERCOME SOIL ACIDITY

There are several ways to mitigate soil acidity, the common practice being liming. Liming with ground magnesium limestone is the standard practice to improve the fertility of the soils, except for rubber and oil palm, which are tolerant to soil acidity. Other methods include organic matter amendment and gypsum application. Most of the studies conducted on soil acidity focus mainly on the liming of topsoils under conventional tillage. Relatively little research effort has been devoted to subsoil acidity and strategies for correcting acidification for cultivation with minimum or no tillage such as grassland or cultivated forests.

Acidity in both top- and subsoils often produces toxic levels of Al and Mn and deficiencies of Ca. Root extension and proliferation are limited by toxic levels of Al and/or deficient levels of Ca with the result that the crop suffers both drought and nutrient stresses because the limited root system is no longer efficient in taking up water and essential elements for growth. These stresses lead to reduced crop yield and quality.

Severe acidification of topsoil can lead to the transfer of acidity to subsoil, which would then be negatively impacted. Once subsoil has been acidified, amelioration by lime incorporation is impractical and expensive. Alternative strategies such as the use of surface applied gypsum are recommended in this situation.

There is no natural deposit of gypsum in Malaysia, and for that reason ameliorating subsoil acidity of Malaysian soils has never been emphasized. However, by-product of gypsum or red gypsum is available in the country, but unfortunately, it is still classified as scheduled waste. Thus, more testing need to be conducted to provide evidence of its usefulness in crop production without adverse effect on the quality of the crops and the environment.

For acid sulfate soils, a large amount of liming material is required to raise the soil pH by one unit. Therefore, the recommended practice is to keep the pyrite layer unoxidized by keeping this layer under submerged condition. For areas where the pyrite layer forms right to the surface, growing flooded rice with proper liming rate is recommended.

The problem for peatland area is the injury caused by proton pressure to the roots of the plants due to low pH. In general, raising the soil pH to above 4.0 will overcome the root injury due to proton pressure. According to Husni et al. (1995), pH, percent humic acid, and buffer capacity are factors influencing the lime requirement of tropical peat.

SOIL ORGANISMS, THEIR BENEFICIAL ACTIVITIES, HABITAT, AND DIVERSITY IN THE SOILS OF MALAYSIA

INTRODUCTION

From the moment a natural system is modified by human activities for agricultural purposes, major changes occur to the soil environment and to the flora and fauna populations and community present. The intensity of the change induced when compared with the original ecosystem and the ability of the various organisms to adapt to these changes will determine the ultimate community present after the perturbation. This community will further be modified as the agricultural practices are altered to suit human needs and changing agricultural paradigms. Practices that are generally considered as beneficial mostly involve the management of organic matter, particularly the control of the quality or quantity of residues added or kept on the soil surface (litter) and the reduction or complete absence of soil disturbance (tillage) (Hendrix et al., 1990). Crop rotation and diversification also play an important role in increasing the diversity of food resources and environmental conditions for the soil biota. Other corrective practices such as fertilization and liming are also important and are generally considered to have positive effects on most organisms (although negative effects may occur under some conditions for some organisms).

SOIL ORGANISMS

Most of the soils contain abundant living organisms that affect soil structure and nutrient cycling. These microorganisms live in the rhizosphere, or the root zone, the area of partnership between plant roots, soil, and soil organisms. Figure 5.4 shows three broad groups of belowground organisms: microfauna, mesofauna, and macrofauna. Microfauna are an enormous, microscopic class that includes protozoa and fungi (primary agents of organic matter decay, bind soil aggregates), actinomycetes

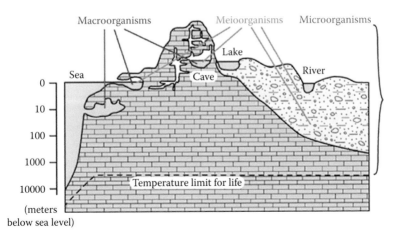

FIGURE 5.4   Soil organisms in the soil. (From Plant & Soil Sciences eLibrary^PRO, Soil genesis and development, lesson 6—Global soil resources and distribution. Accessed December 11, 2014. http://passel.unl.edu/pages/informationmodule.php?idinformationmodule=1130447 033&topicorder=12&maxto=12&minto=1, 2014.)

(decomposers of organic matter, the *smell* of soil), and bacteria (decomposition of organic and inorganic material, fixation of nitrogen). Mesofauna (nematodes and rotifers) help regulate microbial populations.

Normally, soils under agricultural use have a huge number of microfauna and mesofauna. A gram of dry weight soil may contain approximately 10^7–10^9 bacteria, 10^7–10^8 actinomycetes, 10^5–10^6 fungi, 10^4–10^5 protozoa; and 10–100 nematodes. Macrofauna (earthworms, insects) accelerate organic matter decomposition, mix organic matter and soil together, and aerate the soil by channeling and burrowing.

Soil organisms such as insects (e.g., corn root worm) and plant disease pathogens (e.g., seed rotting fungi) could be harmful to crops, but some bacteria (rhizobia) and fungi (mycorrhizae) associated with roots are helpful. Other bacteria and fungi are accountable for vital soil processes such as plant residue degradation and nitrogen mineralization from organic matter. Earthworms are a positive indicator of soil quality and productivity. Reduced tillage systems have more earthworms than conventional tillage systems. Similarly, other beneficial organisms can be promoted through organic practices (Hendrix et al., 1990).

Microorganisms are abundant in soil, but to observe their beneficial effects on plants and environment their population in soil must be increased. Hence, the application of the effective microbes intensified the biological soil activity and improved the physical and chemical soil properties, hence contributing healthier plant growth and development (Lee et al., 2008).

Macroorganisms

More than 20 taxonomic groups of macrofauna could be grouped into a variety of different functional classes, depending on their activity and effects on the soil environment. Representatives of the macrofouna in the soil-surface litter is as shown in Figure 5.5. One of the most important and widely used division is that of beneficial

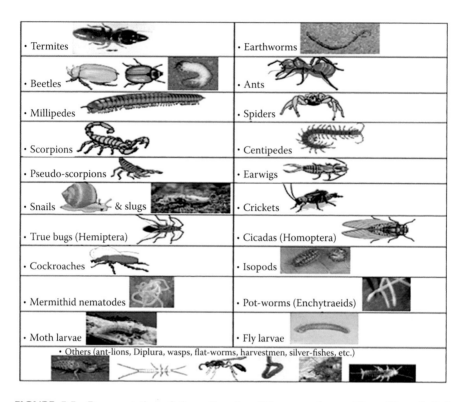

• Termites	• Earthworms
• Beetles	• Ants
• Millipedes	• Spiders
• Scorpions	• Centipedes
• Pseudo-scorpions	• Earwigs
• Snails & slugs	• Crickets
• True bugs (Hemiptera)	• Cicadas (Homoptera)
• Cockroaches	• Isopods
• Mermithid nematodes	• Pot-worms (Enchytraeids)
• Moth larvae	• Fly larvae
• Others (ant-lions, Diplura, wasps, flat-worms, harvestmen, silver-fishes, etc.)	

FIGURE 5.5 Representative of the soil-surface litter macrofauna. (From Plant & Soil Sciences eLibrary[PRO], Soil genesis and development, lesson 6—Global soil resources and distribution. Accessed December 11, 2014. http://passel.unl.edu/pages/informationmodule. php?idinformationmodule=1130447033&topicorder=12&maxto=12&minto=1, 2014.)

and adverse (or pest) soil organisms. Therefore, most of the characterizations of the soil macrofauna communities at a particular site are restricted to identification of better known groups such as earthworms, termites, ants, and beetles, and the complete presentation of the data is generally limited only to taxonomic group level (higher taxa) or morphospecies. Soil fauna are highly variable and they are adaptable to the environment. They can feed on various available food sources ranging from herbivores to omnivores and carnivores. Several by-products of these organisms are used as food resources by other soil fauna.

Population of earthworms can be influenced by several factors. In Malaysia, the earthworm population densities in oil palm are affected by plant age and soil type. Four major factors that dictate the heterogeneity of earthworm population in oil palm plantation are: (1) food and soil physical habitat, (2) exchangeable calcium, (3) pH, and (4) exchangeable potassium (Sabrina et al., 2009). A higher number of earthworms are observed under 7–14-year old compared to >14-year old palms.

Functions of the Macroorganisms

Macroorganisms are the soil-feeding (geophagous) bioturbators, important in opening channels within the soil and on its surface, affecting hydrological processes and gaseous exchanges, as well as modifying soil structure, aggregate formation, and even soil formation rates. Lastly, the predatory organisms that act at the top of the soil food chain are also counted between the beneficials, feeding on other soil and surface dwelling or active organisms, controlling their populations, and often helping to counteract pest outbreaks (thus acting as bio-control agents). Earthworms constitute the largest part of invertebrate biomass in most soils (Tondoh et al., 2007). Soil macrofauna increased with leaf litter calcium and decreased with leaf litter carbon in the plantations (Reich et al., 2005). Although they did not find any differences in soil macrofauna abundance and in biomass of nitrogen fixing tree plantations in comparison with nonnitrogen fixing tree ones, *Acacia salicina* (as a nitrogen fixing tree species) had the highest soil macrofauna abundance and biomass (especially of earthworms) in comparison with other tree plantations, and *Eucalyptus camaldulensis* (as a nonnitrogen fixing tree species) plantation had the lowest.

Soil communities affect on the processing of organic matter and nutrients. Soil faunal activity could improve soil physiochemical properties. The studies in the tropics have proved the vital role of soil fauna in the regulation of plant litter decomposition and nutrient release. Litter-feeding organisms hasten N mineralization in temperate, deciduous woodlands (Anderson et al., 1985). Earthworms constitute the largest part of invertebrate biomass in most soils (Tondoh et al., 2007). The activity of the organisms affects the soil processes that control the availability of plant nutrients such as nitrogen (Zou & Bashkin, 1998), organic matter dynamics and land productivity (Reich et al., 2005; Barrios, 2007). The soil and litter arthropods can be the useful bioindicators of the effects of land management on nutrient dynamics and site productivity (Bird et al., 2004). At the local level, soil properties (Mathieu et al., 2004) and litter quality and quantity (Aubert et al., 2003) are the most important factors that regulate macroinvertebrate communities (Tsukamoto and Sabang, 2005).

Soil macroinvertebrates can be associated with litter quality more than with litter quantity. Tree species rich in calcium were associated with increased native earthworm abundance and diversity, as well as with increased soil pH, exchangeable calcium, percent base saturation, and forest floor turnover rate (Reich et al., 2005).

Biologically, higher plants influence the life of most of the organisms (Antunes et al., 2008). Distribution of earthworms is regulated by leaf litter quality (Ca, C, and N), whereas the macrofauna richness is regulated by leaf litter mass, soil organic carbon, and leaf litter Mg. As the soil macrofauna is an important factor regulating the litter decomposition, more studies should be conducted on the relationship of soil mac- rofauna abundance and richness with litter decomposition, particularly the influence of seasonal variation (Sayad et al., 2012).

One of the earthworm species, *Eisenia fetida* (Figure 5.6) is viewed as a possible alternative of protein source in fish meal. The earthworms are being used as fish

FIGURE 5.6 An earthworm in its burrow mixing the residues on the soil surface into the subsoil. (Courtesy of USDA-NRCS.)

bait, and this practice is common in fishing activity throughout Malaysia. Owing to their high reproductive rate, low feeding costs, and ease of breeding in captivity, earthworms constitute an extremely interesting protein source for fish feed. Having high protein content earthworms are also used to feed chickens and pigs, and as a dietary supplement for ornamental fish and for fish cultivation (Shim and Chua, 1986; Zakaria et al., 2013).

Consumption of the finished product can help in improving soil biological, physical, and chemical properties and hence can improve the soil environmental quality (Ismail 2005). Vermicomposting complicated feeding of the epigeic earthworms with organic waste for the production of vermicast. Epigeic earthworms such as *E. fetida*, *E. andrei*, *Perionyx excavatus*, and *Eudrilus eugeniae* have been used to convert organic waste into vermicast or the worm feces that can be used as organic fertilizer and soil conditioner (Garg et al., 2006).

The nutrient contents of the oil palm biomass fibers were analyzed prior to the vermicomposting process. Due to the presence of earthworms in the vermicomposters, the decomposition was higher than the other treatments as depicted in Figure 5.7. Subsequently, the earthworms can ameliorate the polluting materials and are focused on biological degradation processes (Thambirajah et al., 1995). The application of earthworms in vermicomposting of oil palm empty fruit bunch (EFB) has been shown to increase the heavy metals as earthworms can accumulate a considerable amount of heavy metals in their tissues as shown in Table 5.8. Reduction in weight and volume of organic substrate during vermicomposting may be the reason for the increase in heavy metal concentration in the end product. The palm oil sludge from palm oil mill could be applied as efficient soil conditioner for sustainable land practices, after processing by composting with earthworms (Nahrul Hayawin et al., 2012).

In the natural Malaysian environmental conditions, the capability of *E. fetida* to convert organic waste into vermicast differs with type and quality of the substrate

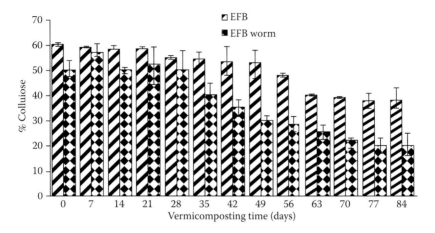

FIGURE 5.7 Trends in empty fruit bunches (EFB) cellulose degradation during vermicomposting (With and without earthworms).

TABLE 5.8

Heavy Metal Content (mg kg⁻¹) in Initial Feed Substrates and Vermicomposts Obtained from Empty Fruit Bunch (EFB) + Palm Oil Mill Effluent (POME) Vermicomposters

Feed Mixtures	Cu	Fe	Zn	Mn
Heavy Metal Content in Initial Mixtures[a]				
V_1	$3 \times 10^{-2} \pm 3 \times 10^{-3}$	$1.6 \pm 1 \times 10^{-3}$	$0.56 \pm 2 \times 10^{-2}$	1.3 ± 0.1
V_2	$0.18 \pm 2 \times 10^{-3}$	18.2 ± 2.3	0.59 ± 0.1^{-2}	1.5 ± 0.2
V_3	$0.18 \pm 6 \times 10^{-3}$	23.9 ± 0.1	$0.79 \pm 1 \times 10^{-2}$	2.1 ± 0.3
V_4	$0.19 \pm 1 \times 10^{-2}$	23.6 ± 0.6	$0.78 \pm 2 \times 10^{-2}$	2.3 ± 0.3
V_5	$0.3 \pm 8 \times 10^{-3}$	25.6 ± 1.2	$0.8 \pm 2 \times 10^{-2}$	2.5 ± 0.1
V_6	$0.4 \pm 2 \times 10^{-3}$	23.9 ± 1	$1.1 \pm 8 \times 10^{-2}$	2.6 ± 0.3
Heavy Metal Content in Final Vermicomposts Obtained from Different Vermicomposters[b]				
(mean + S.E., $n = 3$)				
V_1	$7 \times 10^{-2} \pm 4 \times 10^{-3}$	2.8 ± 1.4	$0.7 \pm 2 \times 10^{-2}$	2.0 ± 0.5
V_2	$0.4 \pm 3 \times 10^{-2}$	27.2 ± 1	$1.4 \pm 3 \times 10^{-2}$	2.8 ± 0.9
V_3	$0.77 \pm 5 \times 10^{-2}$	27.9 ± 1.9	$1.7 \pm 4 \times 10^{-2}$	3.6 ± 0.6
V_4	$0.8 \pm 2 \times 10^{-2}$	28.7 ± 0.9	$1.8 \pm 6 \times 10^{-2}$	4.1 ± 0.5
V_5	0.82 ± 10^{-2}	29.5 ± 0.8	$2.0 \pm 4 \times 10^{-2}$	4.8 ± 0.4
V_6	$1 \pm 3 \times 10^{-2}$	29.8 ± 1	$2.1 \pm 4 \times 10^{-2}$	5.4 ± 1.6

[a] Initial physicochemical characteristics of the feed in the vermicomposters have been calculated based on the percentage of EFB and POME.

[b] Mean value followed by different letters is statistically different (ANOVA; Tukey's test, $P < 0.05$). Units of all the parameters, mg/kg.

Symbols; V_1 = EFB(100), V_2 = EFB(90) + POME(10), V_3 = EFB(80) + POME(20), V_4 = EFB(70) + POME(30), V_5 = EFB(60) + POME(40), V_6 = EFB(50) + POME (50)

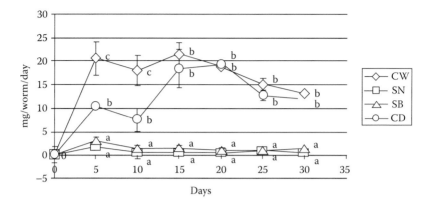

FIGURE 5.8 Growth increment of *E. fetida* given four types of the respective organic wastes (CW, cafeteria waste; SB, shredded banana trunk; SN, shedded newspaper; CD, cow dung).

used. Growth increment of *E. fetida* paralleled with the rate of waste conversion into vermicast, for example, higher growth increment resulted in faster conversion of wastes into vermicast (Jais and Hassan, 2008).

The growth increment of *E. fetida* given different types of substrates is shown in Figure 5.8. The increase in weight occurred immediately after the earthworms were fed with the waste in the order, cafeteria waste > shredded banana trunk > shredded newspaper > cow dung. In general, exponential increase in weight of the earthworm in all the organic wastes tested occurred until the 5th day of vermiculture and declined after the 15th day.

Furthermore, the use of earthworms in manure management has enhanced extremely in recent use of organic wastes, such as crop residues, animal manure, biosolids, and industrial waste (Edwards, 1998). Vermicompost is a distinctive organic manure source due to its plentiful amounts of nutrients, growth-enhancing substances, and many favorable microbes, which include P solubilizing and cellulose decomposing organisms (Sultan, 1997). The vermicomposting has increased the solubility of phosphate rock. The extractable P was 17% higher in vermicompost with the addition of phosphate rock. In addition, extractable macronutrients N and K were also found to be significantly higher in vermicomposting with the addition of phosphate rock (Wei et al., 2012).

Many species of nematodes are well known as important and devastating parasites of humans, domestic animals and plants. Nevertheless, most species are not pests, they occupy any niche that provides an available source of organic carbon in marine, freshwater and terrestrial environments. There may be 50 different species of nematodes in a handful of soil and millions of individuals can occupy 1 m². The nematodes that do not feed on higher plants may feed on fungi or bacteria and others are carnivores or omnivores.

Nematodes (Figure 5.9) have the potential to respond rapidly to disturbance and enrichment of their environment; increased microbial activity in soil leads to changes

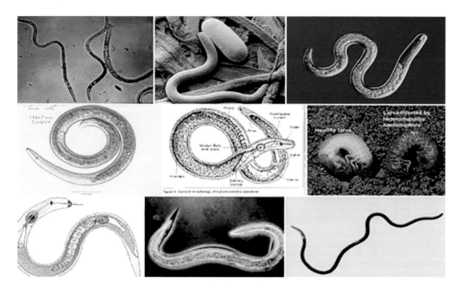

FIGURE 5.9   Beneficial nematodes in the soil. (From Plant & Soil Sciences eLibrary[PRO], Soil genesis and development, lesson 6—Global soil resources and distribution. Accessed December 11, 2014. http://passel.unl.edu/pages/informationmodule.php?idinformationmodu le=1130447033&topicorder=12&maxto=12&minto=1, 2014.)

in the proportion of opportunistic bacterial feeders in a community. Over time, the enrichment opportunists are followed by more general opportunists that comprise fungal feeders and different genera of bacterial feeders (Bongers and Ferris, 1999). This succession of nematode species performs an important role in decomposition of soil organic matter, mineralization of plant nutrients, and nutrient cycling (Hunt et al., 1987).

Bacterial-feeding nematodes have a higher carbon:nitrogen (C:N) ratio (± 5.9) than their substrate (± 4.1) (Ferris et al., 1997), so that in consuming bacteria they take in more N than necessary for their body structure. The excess nitrogen is excreted as ammonia (Rogers, 1989). The C:N ratio of fungal-feeding nematodes is closer to that of their food source. However, for nematodes of both feeding habits, a considerable proportion of the C consumed is used in respiration (perhaps 40% of the food intake) (Ingham et al., 1985). The N associated with respired C that is in excess of structural needs is also excreted. The excreted N is available in the soil solution for uptake by plants and microbes. Because microbivorous nematodes exhibit a wide range of metabolic rates and behavioral attributes, the contribution of individual species to nitrogen cycling and soil fertility may vary considerably.

Soil nematode communities may also provide useful indicators of soil condition. Nematodes vary in sensitivity to pollutants and environmental disturbance. Recent development of indices that integrate the responses of different taxa and trophic groups to perturbation provides a powerful basis for analysis of faunal assemblages in soil as *in situ* environmental assessment systems. Application of nematode faunal composition analysis provides information on succession and changes in decomposition pathways in the soil food web, nutrient status and soil fertility, acidity, and the effects of soil contaminants (Bongers and Ferris, 1999).

SOIL MICROORGANISMS

The natural soils comprise huge populations of microscopic plants and animals present in a state of dynamics equilibrium and changing balances. Microorganisms perform a vital role in agriculture by supplying nutrients to the plants and lessen the demand of chemical fertilizers (Cakmakci et al., 2006). Several bacteria, fungi, and actinomycetes are prominent to execute for the plant growth and have abundant quantity in the soil. Bacteria are more effective in phosphate solubilization than fungi (Alam et al., 2002). Bacteria have potential to be associated with the plants in roots by adhering at its rhizosphere or endosphere regions. Figure 5.10 shows Transmission Electron micrographs of some beneficial bacteria isolated from the wetland, aerobic, and acid sulfate rice areas in Malaysia. Bacterial strains isolated from maize soils are found to produce phytase enzyme important in releasing phosphorus for organic material (Hussin et al., 2010). One of the important fungi is the mycorrhiza which is the symbiotic association between fungi and vascular host plants. Mycorrhizal fungi colonize the root systems of several plants and aid in the uptake of nutrients, thereby improving plant growth and overall health. The beneficial effects of mycorrhiza on growth of several tropical crops have been highlighted by Naher et al., (2013). *Glomus mosseae* is well known to colonise

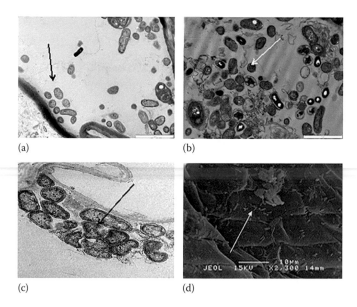

(a)   (b)

(c)   (d)

FIGURE 5.10 Transmission electron microscopic micrographs showing (a and b) nitrogen-fixing bacteria isolated from wetland rice at Tanjung Karang, Selangor, (c) PSB isolated from aerobic rice at Kepala Batas, Penang (d) acid sulfate bacteria isolated from acid sulfate soils at Semerak, Kelantan, Malaysia.

several crops and assist the plants to efficiently uptake nutrients and increase the plant's tolerance to diseases and other stresses. The soil bacteria and fungi form relationships with plant roots that provide important nutrients such as nitrogen and phosphorus. Fungi can colonize the plants and can supply numerous benefits, including drought and heat tolerance, and resistance to insects and plant diseases. Soil properties, vegetation, and fertilizer usage can influence the distribution and population of soil microorganisms.

FUNCTIONS AND BENEFICIAL CHARACTERS OF THE SOIL MICROORGANISMS

RELEASING NUTRIENTS FROM ORGANIC MATTER

Soil microorganisms are accountable for most of the nutrient release from organic matter. When microorganisms decompose organic matter, they use the carbon and nutrients in the organic matter for their own growth. They release extra nutrients into the soil where they can be taken up by plants. If the organic matter has a low nutrient content, microorganisms will take nutrients from the soil to meet their requirements. For example, applying organic matter with C:N ratios lower than 22:1 to soil generally increases mineral nitrogen in soil. In contrast, applying organic matter with C:N ratios higher than 22:1 usually results in microorganisms taking up mineral nitrogen from soil (Hoyle et al., 2011).

FIXING ATMOSPHERIC NITROGEN

Biological nitrogen fixation is an important source of nitrogen for agriculture and may account for up to 80% of total nitrogen inputs (Unkovich, 2003). In symbiosis, rhizobia or bradyrhizobia fix nitrogen gas from the atmosphere and make it available to the legume. In exchange, they receive carbon from the legume. The symbiosis is highly specific, and particular species of rhizobia and bradyrhizobia are obligatory for each legume. Fixation of N can also be occurring in the non-symbiotic association by several bacteria living in the rhizosphere.

Legumes

Legumes are the two most significant flowering plants used in agriculture. Legumes are benefitted as human and animal food, as wood, and as soil-improving components of agricultural and agroforestry systems. In Malaysia only a few legumes are being grown as agricultural crops and limited *Rhizobium* inoculants are being applied. Many forest trees are legumes, which nodulate with the bacterium *Rhizobium* (fast growing) or *Bradyrhizobium* (slow growing) and fix gaseous nitrogen thus using some of the 84,000 tonnes of nitrogen gas in the air above each hectare of land. There are more than 18,000 species of legumes of which about 7,200 species are woody. Only about 18% of these woody species have been inspected for nodulation and of this 92%–94% of the mimosoids and papilionoids nodulated, but only about 34% of the caesalpinoids form nodules (Dobereiner, 1993). It is not easy to determine if tree legumes nodulate as nodules are both difficult to find in forest soils and difficult to assign to a particular tree. Therefore, observations on young plants are helpful to assess the nodulation status of the plant.

Symbiotic Nitrogen Fixation

Most of the soils contain several types of *Rhizobium*, and in some soils, populations of suitable strains may be absent or too small for nodulation formation. In this situation the response to nodulation inoculation with *Rhizobium* might not be expected. Therefore superior strains to be used as inoculants need to be selected. The process of selection usually starts with collection of strains by isolation from nodules of a particular legume under consideration. This is followed by an assessment of their ability to fix nitrogen in a strain trial in pots using a rooting medium that does not comprise rhizobia (Trinick, 1980). For example, root nodules formed on the root system of a soybean plant. Nitrogen-fixing root nodule bacteria (*Bradyrhizobium*) present inside the nodule provide valuable organic nitrogen to the host plant, which promotes plant growth.

Nonsymbiotic Nitrogen Fixation

Biological nitrogen fixation also occurs through nonsymbiotic bacteria growing on roots and in degrading litter; through blue–green algae or cyanobacteria on soil and plant surfaces; and through associations of cyanobacteria with fungi and lichens, or with higher plants such as liverworts, mosses, cycads, and the angiosperm *Gunnera*. Most of the nitrogen in forest ecosystems is derived from biological nitrogen fixation. These systems are very efficient in recycling nitrogen that is leached to lower depths in the soil through uptake by deep roots, and through leaf fall, concentrating this nitrogen in the litter and upper soil horizons. Disturbing this natural cycle, which conserves scarce nutrients so effectively can lead to rapid loss of soil fertility. Maintenance of the litter layer as a soil mulch to reduce erosion as well as to conserve nutrients is a very important aspect of maintaining soil fertility around trees and shrubs.

Diazotrophs Associations in Rice

Diazotrophs are N_2-fixing bacteria that colonize and contribute biological nitrogen to the crops (Kundu and Ladha, 1995). Rice plant can form natural associations with various N_2-fixing bacteria, both phototrophs and heterotrophs. These diazotrophs can improve the growth and development of rice plant by transferring fixed N_2 or by producing phytohormone. The N_2 fixed by asymbiotic diazotroph may not be immediately available for plant growth. The plant may benefit from asymbiotic N_2 fixation in the long term, as nitrogen gets released through biomass turnover (Dobbelaere et al., 2003). Endophytic diazotroph can supply nitrogen more efficiently to the plants. The endophytic association is competitively accomplished to occupy associable niches within this nutrition-enriched and protected habitat of the root interior without showing any pathological symptoms on the host plant (Cocking, 2003). For example, the colonization of wheat roots by strains of *Azospirillum*, a bacterial inoculant that acts as a phytosimulator. Moreover *Azospirillum* induces the proliferation of plant root hairs, which can result in improved nutrient uptake.

Phosphatic fertilizers applied to low pH soil are precipitated by complexes with aluminium and iron immediately after application and making them not available to plant. Besides correcting the soil pH, another alternative to overcome this problem is through the use of phosphate- solubilizing bacteria.

Phosphate-Solubilizing Bacteria

The quantity of phosphate-solubilizing bacteria (PSB) is more abundant in the rhizosphere than nonrhizosphere soil and is metabolically more dynamic than from other sources (Vazquez et al., 2000). The PSB are found everywhere in the soils with different forms and their populations. The population of PSB is affected by various soil properties such as physical and chemical properties, organic matter, P content, and cultural activities (Kim et al., 1998), whereas higher populations of PSB are found in agricultural and rangeland soils (Yahya and Azawi, 1998).

It has been found that the poorly soluble P is usually dissolved by microorganisms, which can then be converted into soluble forms by the process of acidification, chelation, and exchange reactions (Chung et al., 2005). Microorganisms, especially PSB and arbuscular mycorrhizal (AM) fungi have the ability to solubilize P in soil and reduce inputs of chemical fertilizers (Arpana and Bagyaraj, 2007). Species of *Pseudomonas*, *Bacillus*, *Rhizobium*, *Burkholderia*, *Achromobacter*, *Agrobacterium*, *Micrococcus*, *Aereobacter*, *Flavobacterium*, and *Erwinia* are the plant growth-promoting rhizobacteria that have the ability to solubilize insoluble inorganic phosphate compounds, such as tricalcium phosphate, dicalcium phosphate, hydroxyapatite, and phosphate rock (Goldstein, 1986). Several PSB isolates from aerobic rice in Kepala Batas, Penang, Malaysia, are able to solublize insoluble P by producing organic acids (Panhwar et al., 2012). Phosphate solubilizing fungi have also been shown to have the ability to convert insoluble phosphatic compounds into soluble P form. These microorganisms may compensate higher fertilizer cost and also may mobilize the fertilizers added to soil (Pradhan and Sukla, 2005).

ACTINOMYCETES

Actinomycetes are numerous and widely distributed in soil and are next to bacteria in abundance. They are widely distributed in the soil (Fierer et al., 2009). Actinomycetes are fungi-like bacteria creating long filaments that stretch through the soil. They have sometimes been classed as fungi because they both look similar and decompose similar material as fungi. However, they do not have defined nucleus. Also, antibacterial agents work against them but antifungal agents do not. As a decomposer, the actinomycetes specialize in breaking down tough cellulose and lignin found in wood and paper and the chitin found in the exoskeletons of insects. The breakdown of these materials makes nutrients once again available to plants.

The population of actinomycetes increases with depth of soil even up to horizon *C* of a soil profiler. They are heterotrophic, aerobic, and mesophilic (25°C–30°C) organisms, and some species commonly present in compost and manures are thermophilic growing at 55°C–65°C temperature (e.g., *Thermoactinomyces*, *Streptomyces*).

Actinomycetes degrade/decompose all sorts of organic substances such as cellulose, polysaccharides, protein fats, organic acids, and so on. Organic residues/substances added soil is first attacked by bacteria and fungi and later by actinomycetes, because they are slow in activity and growth than bacteria and fungi. They decompose/ degrade the more resistant and indecomposable organic substance/matter and produce a number of dark black to brown pigments, which contribute to the dark color of soil humus. They are also responsible for subsequent further decomposition of humus (resistant material) in soil. They are not only able to survive under extreme soil condition such as low level of moisture or high salinity, but actinomycetes are also reported to promote plant growth (Hamdali et al., 2008). Actinomycetes are one of the predominant members of soil microbial communities, and they have beneficial roles in soil nutrients cycling and agricultural productivity (Elliot and Lynch, 1995).

Degrading Pesticides

The degradation of agricultural pesticides in the soil is primarily performed by microorganisms. Some microorganisms in soil produce enzymes that can break down agricultural pesticides or other toxic substances added to the soil. The length of time these substances remain in soil is related to how easily they are degraded by microbial enzymes.

Controlling Pathogens

Microorganisms and soil animals infect plants and decrease plant yield. However, many organisms in the soil control the spread of pathogens. For example, the occurrence of some pathogenic fungi in soil is decreased by certain protozoa that consume the pathogenic fungi. The soil food web contains many relationships similar to this that decrease the abundance of plant pathogens.

Improving Soil Structure

Biological processes in soil can develop soil structure. Some bacteria and fungi produce substances during organic matter decomposition that chemically and physically bind soil particles into microaggregates. The hyphal strands of fungi can cross-link soil particles helping to form and maintain aggregates. A single gram of soil can contain several kilometers of fungal hyphae (Young and Crawford 2004). In addition, soil animals increase pores by tunneling through soil and increase aggregation by ingesting soil.

ECOLOGICAL SIGNIFICANCE OF SOIL MICROORGANISMS

Soil microorganisms are very important as almost every chemical transformation taking place in the soil involves active contributions from soil microorganisms. In particular, they play an active role in soil fertility as a result of their involvement in the cycle of nutrients such as carbon and nitrogen, which are required for plant growth. For example, soil microorganisms are responsible for the decomposition of the organic matter entering the soil (e.g., plant litter) and therefore in the recycling of nutrients in soil. Certain soil microorganisms such as mycorrhizal fungi can also increase the availability of mineral nutrients (e.g., phosphorus) to plants. Other soil microorganisms can increase the amount of nutrients present in the soil. For instance, nitrogen-fixing bacteria can transform nitrogen gas present in the soil atmosphere into soluble nitrogenous compounds that plant roots can utilize for growth. These microorganisms, which improve the fertility status of the soil and contribute to plant growth, have been termed *biofertilizers* and are receiving increased attention for use as microbial inoculants in agriculture. Similarly, other soil microorganisms have been found to produce compounds (such as vitamins and plant hormones) that can improve plant health and contribute to higher crop yield. These microorganisms (called *phytostimulators*) are currently studied for possible use as microbial inoculants to improve crop yield.

BIODIVERSITY AND HABITAT

Each animal, plant, and microbe species requires a slightly different habitat. Thus, a wide variety of habitats are required to support the tremendous biodiversity on earth. At the microbial level, diversity is beneficial for several reasons. Many different organisms are required in the multistep process of decomposition and nutrient cycling. A complex set of soil organisms can compete with disease-causing organisms and prevent a problem-causing species from becoming dominant. Many types of organisms are involved in creating and maintaining the soil structure that is important to water dynamics in soil. Many antibiotics and other drugs and compounds used by humans come from soil organisms. Hence, preserving the diverse and healthy ecosystem is crucial. (Pankhurst, 1997).

The large quantity of living things on Earth requires a few basic elements: air, food, water, and a place to live. The decomposers in soil have need of an appropriate physical environment or habitat to do their work. All soil organisms require water for their activities and in condition of less water, some of them can still survive for long periods by transforming into a spore-like structure. A majority of the living organisms are aerobic requiring oxygen for growth, though some have evolved to thrive when oxygen is absent (anaerobes). Greater soil porosity and a wide range of pore sizes in the soil allow these organisms to get the oxygen. Soil texture has a great influence on the available habitat for soil organisms. Finer soils have a larger number of small micropores that provide habitat for the microorganisms. In addition, all soil organisms require organic material to use as energy and carbon source. Those requiring complex carbohydrates are known as the heterotrophs and those require carbon from carbon dioxide and energy from the inorganic materials are known as the autotrophs. Supply of fresh organic materials to soil will stimulate vigorous growth of soil organisms. Soil supports the growth of a variety of unstressed plants, animals, and soil microorganisms, usually by providing a diverse physical, chemical, and biological habitat. The ability of soil to support plant and animal life can be assessed by measuring the following indicators: Biological activity indicators include active fungi, earthworms, microbial biomass, potentially mineralizable nitrogen, respiration, and soil enzymes. Biological diversity indicators include habitat diversity and diversity indices for organisms such as bacteria, macro and microarthropods, nematodes, and plants.

The importance of knowledge in biological diversity provides opportunities in biotechnology and commercialization. Soil is full of living organisms of various sizes, ranging from large, easily visible plant roots and animals, to very small mites and insects, to microscopically small microorganisms (e.g., bacteria and fungi.) Microorganisms are the primary decomposers of the soil and perform processes such as transforming and recycling of organic materials, thereby mineralizing nutrients for growth of new plants and organisms.

The plant growth-promoting rhizobacteria from different states of Malaysia (Kedah, Kelantan, Selangor, and Terengganu) have potential for the multiple beneficial characteristics such as N2 fixation, P solubilization, K solubilization, indole acetic acid (IAA), and enzyme production (Tan et al., 2014a). Several N-fixing bacterial species have been isolated from rice growing area in Tanjung Karang, Selangor (Naher et al.,2009). The bacterial strains able to carry out N fixation are as presented in Table 5.9. Panhwar et al. (2012) isolated phosphate solubilizing bacteria from aerobic rice grown at Kepala Batas, Penang, Malaysia, having the ability to produce different organic acids for phosphate solubilization as presented in Table 5.10.

TABLE 5.9

Nitrogen-Fixing Bacteria, Their habitat and Mechanisms of Crop Improvement

Bacterial Strain	Habitat	Energy Source	Mechanism of Effect	References
Azospirillum spp.	Rhizosphere, midly endophytic in roots, stems and leaves	Root exudates and plant tissue Organics in soil	BNF, PGP	Reinhold and Hurek (1988) Mirza et al. (2000)
H. seropedicae	Endophytic rhizosphere	Root exudates	BNF, PGP	Baldani et al. (2000)
Azoarcus sp.	Endophytic	Root exudates	BNF	Hurek et al. (1994)
B. vietnamiensis	Rhizosphere, endophytic	Root exudates Organics in soil	BNF, PGP	Baldani et al. (2000)
R. leguminosarum bv. Trifolii	Endophytic in roots	Root exudates	BNF, PGP	Biswas et al (2000)
R. etli bv. Phaseoli	Endophytic in roots	Root exudates	PGP	Guttiérrez-Zamora and Martínez-Romero (2001)
Rhizobium and *Corynebacterium* spp.	Endophytic in roots	Root exudates	BNF, PGP	Naher et al. (2009)
Bacillus sp. (Sb 42)	Endophytic in roots	Root exudates	BNF, PGP	Mutalib et al. (2012)
Rhizobium sp. *Bradyrhizobium* sp. *Bacillus* sp.	Endophytic in roots	Root exudates	BNF, PGP	Tan et al. (2014)

Source: Kennedy, I.R. et al., *Soil Biol. Biochem.*, 36, 1229–1244, 2004; Naher et al. (2009); Tan, K.Z. et al., *Am. J. Agric. Biol. Sci.*, 9, 342–360, 2014a.
BNF, biological nitrogen fixation; PGP, plant growth promotion.

TABLE 5.10

Productuion of Organic Acid by Phosphate-Solubilizing Bacteria

Bacterial Species	Organic Acids Produced	Reference
Bacillus sp. (PSB1)	Oxalic, citric acid, succinic acid, malic acid	
Bacillus sp. (PSB6)	Oxalic, citric acid, succinic acid, malic acid	
Bacillus sp. (PSB9)	Oxalic, citric acid, succinic acid, malic acid	
Bacillus sp. (PSB10)	Oxalic, citric acid, succinic acid, malic acid	Panhwar et al. (2012)
Bacillus sp. (PSB14)	Oxalic, citric acid, succinic acid, malic acid	
Bacillus sp. (PSB15)	Oxalic, citric acid, succinic acid, malic acid	
Bacillus sp. (PSB16)	Oxalic, citric acid, succinic acid, malic acid	

Source: Panhwar, Q.A., *African J. Biotechnol.*, 11, 2711–2719, 2012.

The microbial activities were conducted at forest plantation in Sabah, Malaysia. The soil disturbance affected the microbial activities severely, but slowly it was improved after 2 years. Hence, microbial growth kinetics were proven to be the promising tools for assessing the effect of soil disturbance and rehabilitation (Ilstedt, 2002). Moreover, a study was conducted at Sungai Buloh, Selangor, and Malaysia to observe the effects of herbicides (alachlor and metachlor) on the microbial population especially on the bacterial and fungal population. It was found that application of alachlor has less effect on bacterial population when compared with metachlor. It shows that the herbicide application significantly influences the microbial activities in the soil (Ismail and Shamsuddin, 2005).

Soil bacterial communities of tropical rainforest in Malaysia are affected by the environmental distance that was highly correlated with community dissimilarity at both spatial scales, stressing the greater role of environmental variables rather than spatial distance in determining bacterial community variation at different spatial scales. Soil pH was the only environmental parameter that significantly explained the variance in bacterial community at the local scale, whereas total nitrogen and elevation were additional important factors. In total, our results support a strong influence of the environment in determining bacterial community composition in the rainforests of Malaysia. (Tripathi et al., 2014). Similar results were found earlier by that bacterial community composition, and diversity was strongly correlated with soil properties, especially soil pH, total carbon, and C/N ratio. Soil pH was the best predictor of bacterial community composition and diversity across the various land-use types, with the highest diversity close to neutral pH values (Tripathi et al., 2012). Moreover, the herbicide application to soil of oil palm plantation causes transient impacts on microbial population growth, when applied at recommended rate or even as high as double (2×) of the recommended field application rate (Zain et al., 2013).

Many earlier studies of Malaysian fungi were of a floristic nature with the most important work being contributed by the late E.J.H. Corner (Watling and Ginns, 1998). Chin (1988) reported the edible and poisonous fungal species of Sarawak, in particular those used by the indigenous Iban, Melanau, and Malays. The edible and medicinal species of macrofungi are listed in Table 5.11.

TABLE 5.11
Edible and Medicinal Species of Macrofungi

Source	Edible Species	Medicinal Species
Burkill (1966)	21	11
Chin (1981, 1988)	50	–
Chang (1997)	8	9
Chang and lee (2001)	4	3
Total number of species	83	23
Total number of different species	71	12

There is a gradual loss of knowledge in traditional wild edible fungi in the world including Malaysia. A documentation of known wild edible fungi in Malaysia was reported by Abdullah and Rossea (2009) and among those that have been successfully documented were *Cookeina* and *Galiella* of the class *Ascomycetes* and *Termitomyces, Schizophyllum, Hygrocybe, Lentinus, Calvatia, Calostoma* and *Auricularia*, from the *Basidiomycetes* that make up a total of 13 or 14 species belonging to 9 genera. Other reports on utilization of macrofungi by local communities in Peninsular Malaysia also have been reported (Chang and Lee, 2001).

Microbial activities can be observed in all soil types. In forest soil, differences can be observed in microbial population and biomass carbon between a natural forest and an 18-year-old stand of *Shorea leprosula* in Chikus Forest Reserve, Perak, Malaysia (Daljit Singh et al., 2011). Several fungal strains from different locations in the state of Perak and National Park in Pahang, Malaysia with potential antimicrobial activities have also been observed (Siti Hajar, 2011).

Actinomycetes are widely distributed in different habitats and they are involved in important processes such as composting and acts as a biocontrol agent for plant diseases. Several actinomycetes have been isolated from soil in Serdang, Bangi, Petaling Jaya, and Putrajaya areas showing the ability to degrade cellulose, mannan, and xylan components. A few of them have the ability to produce antimicrobial activity against selected phytopathogens such as *Xanthomonas campestris* (Jeffrey et al., 2007).

MICROBES IN AGRICULTURE AND ENVIRONMENT

Microorganisms in soils are the important component that helps to improve agricultural productivity. Naturally-occurring organisms are used to develop biofertilizers and biopesticides to assist plant growth and to control weeds, pests, and diseases. Microorganisms that live in the soil actually help plants to absorb more nutrients. Plants and these friendly microbes are involved in nutrient recycling. The microbes help the plant to take up essential energy sources. In return, plants release organic compounds into the rhizosphere zone for the microbes to use as their carbon and energy sources. Scientists use these friendly microorganisms to develop biofertilizers. The broad application of microbes in sustainable agriculture is due to the genetic dependency of plants on the beneficial functions provided by symbiotic cohabitants (Noble and

Ruaysoongnern, 2010). The agronomic potential of plant–microbial symbioses proceeds from the analysis of their ecological impacts, which have been best studied for N_2 fixing (Franche et al., 2009). This analysis has been based on applied coevolutionary research (Arnold et al., 2010), addressing the ecological and molecular mechanisms for mutual adaptation and parallel speciation of plant and microbial partners.

The major impact of agricultural microbiology on sustainable agriculture would be to substitute agrochemicals (mineral fertilizers, pesticides) with microbial preparations. However, this substitution is usually partial and only sometimes may be complete. Improvement of the legumes symbioses for increased N_2 fixation is crucial. Importance is given to breeding of the leguminous crops for the preferential nodulation by highly active rhizobia strains, for the ability to support N_2 fixation under moderate N fertilization levels and to ensure a sufficient energy supply of symbiotrophic nitrogen nutrition (Provorov and Tikhonovich, 2003). This approach is most promising in legume–rhizobia symbioses where the strong correlations between the ecological efficiency of mutualism and its genotypic specificity are evident (Provorov and Vorobyov, 2010).

The extreme erosion of topsoil from farmlands caused by intensive tillage and row crop production may cause extensive soil degradation that contributed to the pollution of both surface and groundwater. The organic wastes from animal production, agricultural and marine processing industries, and municipal wastes (i.e., sewage and garbage), have become major sources of environmental pollution in both developed and developing countries. In addition, the production of methane from paddy fields and ruminant animals and production of carbon dioxide from the burning of fossil fuels, land clearing, and organic matter decomposition have been linked to global warming as greenhouse gases (Parr and Hornick, 1992). Chemical-based conventional system of agricultural production is one of the major causes to create many sources of pollution that, either directly or indirectly, can contribute to degradation of the environment and destruction of our natural resource base. This situation would change significantly if these pollutants could be used in agricultural production as sources of energy.

Therefore, it is necessary that future agricultural technologies should be compatible with the global ecosystem and with the solutions to such problems in areas different from those of conventional agricultural technologies. An area that appears to hold the greatest promise for technological advances in crop production, crop protection, and natural resource conservation is that of beneficial and effective microorganisms applied as soil, plant, and environmental inoculants (Higa, 1995).

DETRIMENTAL EFFECTS OF ORGANISMS IN SOIL

Highly specialized interactions between soil pathogen and plants can adversely affect seedlings and even adult trees. Several organisms target younger plants but others appear as problems at later stages in the life of the plant. Other pathogens are able to cause disease in many different plant species. Some plant pathogens depend on their host plant for survival and are unable to complete their life cycle without infecting their host plant. Biotrophic organisms of this type are often difficult to grow in laboratory media.

Insect pests are a problem in agriculture production. They attack by defoliating, sucking, stem-mining, and gall-forming species can delay seed ripening, reduce seed production and individual seed weights, reduce the rates of shoot and root growth, enhance the susceptibility of plants to disease, and decrease the competitive potential of plants relative to their unattacked neighbors (Crawley, 1989). Termites are one of the major causes for low yield of the several crops. These insects mostly attack on the roots of the plants and finally cause death of the plants. Method of controlling these insects through agrofriendly techniques such as utilizing some plants as a cheap source of natural pesticides against termites is of importance (Thamer, 2008).

Nematodes such as *Meloidogyne incognita* and many others are associated with plant disease and is a serious threat to crop production worldwide. Several species have been known to cause damage to fruit crops such as guava, banana, and other crops such as chilli, black pepper, and turf grass. Kenaf (Hibiscus cannabinus) cultivated in Telaga Papan in Terengganu, Malaysia has been observed to be affected by the root-knot nematode resulting in reduced plant growth with symptoms of decoloration, drying, and wilting of leaves along with development of galls on roots (Tahery et al., 2011). Population and distribution of this plant parasitic nematodes have been surveyed in banana plantations in various states in the Peninsula Malaysia and found that several nematode species exist in different locations of banana plantations (Sayed Abdul Rahman et al., 2014).

CONCLUSION

The soils of Malaysia contain huge populations of micro- and macroorganisms that are present in a state of dynamics equilibrium and changing balances. These organisms need to be increased in quantity and functionality to benefit the soil productivity. The major impact of these soil organisms on sustainable agriculture would be to substitute agrochemicals with microbial inoculants. The combination of a diversity of biological activities and functions and natural and regulatory mechanisms in the soil forms long-term sustainability. More attention should also be focused on evaluating the role of specific groups of soil organisms, their diversity on the multiple biological interactions in soil to reduce the chemical (herbicides, insecticides, and fertilizers) dependence, and their detrimental effects on humans and the environment.

MECHANICAL PROPERTIES OF SOILS BASED ON ENGINEERING CHARACTERISTICS

Importance of Soils in Construction

Soils are used as bedding or support for all types of heavy structures, such as roads, highways, infrastructure and for foundation of structures. The deterioration or collapse of these structures actually depends on soil behavior, that is, strength parameters of soil. Structural response of soils are expressed in terms of stress, strain, and deflection, which depend on base soil (Brown, 1996). The desirable properties of foundation soil are as follows:

1. Adequate shear strength
2. Adequate permeability
3. Ease and permanency of compaction
4. Volume stability
5. Permanency of strength

Strength parameters of materials are expressed in terms of California bearing ratio (CBR), which was developed in the year 1930. CBR of foundation soil plays the most vital role for structures in terms of durability. CBR of soil fluctuates with variation in water table or flood water. Support provided by the soil in place (subgrade) is the most basic factor for all structural design procedures. Surface deflection of roads is highly dependent on subgrade support.

Some of the available soil parameters are very old. Proctor and modified Proctor test, maximum dry density (MDD), and optimum moisture content (OMC) were developed around the year 1933. Field density test (FDT) was also developed at that time with Proctor tests. Presently, all these soil parameters are used, but they are not reliable. The resilient modulus of soil is measured in the laboratory for MDD and OMC to reflect the conditions under which the subgrade soil is usually prepared. Presence of water in subgrade soil is limited by OMC and controls the strength of soils.

Water is a polar molecule and plays an important role in all types of structures as it determines the durability of structures. Many tests are involved in soil characterization. They are as follows:

1. Innovative soil classification (USC)
2. Consistency of soil-Atterberg limits (McBride, 2002)
3. FDT
4. Resilient modulus test, Mr
5. Resistance value, R
6. Modulus of subgrade reaction, K
7. The standard Casagrande soil test (1932, 1948)
8. CBR
9. Dynamic cone penetration (DCP)
10. Hveem resistance value
11. Plate load test
12. Triaxial tests

Support provided by the soil in place (subgrade) is the most basic factor for all structural design procedures. Surface deflection is highly dependent on subgrade support.

AVAILABILITY OF SOILS IN MALAYSIA

Soils are of low cost and light weight construction materials, which have gained popularity in building industry. Peat is one of the major soils in Malaysia. About 3.0 million ha or 8% of the area is covered with peat. It is located in all over the earth except in the arctic and desert areas and amounts to about 30 million ha or 5% to 8% of the earth's surface (Harden and Taylor, 1983). Two-thirds of the earth coverage of tropical peat is in South East Asia, that is, about 23 million ha. Peat soils encompass 2,457,730 ha; 7.45% of Malaysia's total land mass (32,975,800 ha) (Huat et al., 2005). Sarawak supports the largest area of peat soils in Malaysia, about 1,697,847 ha; 69.08%, Peninsular Malaysia about 642,918 ha; 26.16%, Sabah about 116,965 ha; 4.76% (Wetlands International—Malaysia Ministry of Natural Resources and Environment, 2012). Organic and peat soil are more difficult to stabilize due to lower solid content, higher water content, lower pH, and its potential to interfere chemically and biologically with time and environmental condition (Hernandez Martinez and Al Tabbaa, 2009; Huat, 2002). Laterite soils are also available in many locations of Malaysia. Its CBR is as high as 80%, whereas the required CBR of soil is greater or equal to 5% for major infrastructure construction.

LANDUSE IN MALAYSIA: AN ENGINEERING PERSPECTIVE

About 72% of Malaysia's land is still forested or marshlands according to 1990 data. The remaining land (about 28%) is mostly utilized for agricultural cultivation of palm oil (*Elaeis guineensis*), rubber (*Hevea brasiliensis*), cocoa (*Theobroma cacao*), coconut, and paddy. In Peninsular Malaysia, the general pattern of landuse is one of extensive cultivation in areas where the terrain is less rugged and easily accessible. A large part of the steep mountainous area is comprised of forest. Pockets of steep areas in the lowlands have been opened up for large-scale agricultural development schemes. In Peninsular Malaysia, about 5000 ha of steep mountainous land in the Cameron Highlands has been developed for plantation of tea, temperate vegetables, and fruit trees. In Sabah and Sarawak, much of the cocoa and pepper are planted on steep lands. Other landuse in steep areas is now shifting from cultivation mainly found in East Malaysia. In Sarawak, this amounts to about 0.08 million ha (or 2.7 million ha, including fallow area). The usage of lateritic soils that are found in less steep areas is more extensive despite their stony nature, indicating that the limitations of stony soil are more acceptable to the farmers (Aminuddin et al., 1990). The strength of laterite soils that are expressed in terms of CBR possesses higher strength and is used as construction materials. In general, laterite possesses 80% CBR, which is why it is used as construction material. In many parts of Malaysia, laterite is used for construction of road shoulder such as Rawang, Assam Jawa in state of Selangor.

As landuse is devoted to human activities, it will affect the surrounding areas. Changes in landuse can cause transformation in surface runoff, flood frequency, base flow, and annual mean discharge of water as well as deforestation (Huntington, 2006). Clearing of land plays an important role in soil degradation and loss of soil in catchment areas throughout the world (Gharibreza et al., 2013).

The combined impacts of changing large swaths and climate variability have resulted in increased surface runoff, water yield, soil water content, and evaporation. Moreover, it decreases groundwater flow and percolation. These findings show that the variation of landuse plays a vital role in local water cycle changes, especially for the water movement within the soil layers (Tan et al., 2014b).

IMPACT OF LANDUSE IN MALAYSIA

The topography of Peninsular Malaysia is dominated by the main range, which runs almost centrally along the middle of the Peninsula to a height of about 2000 m above mean sea level. From these mountainous systems, many rivers flow toward the flood plains and the coast. The west coast is dominated by alluvial marine deposits, whereas the east coast has exposed riverine deposits and sandy beach ridges. About 8% of the land area is swamp land, mainly in the coastal depressions. Sarawak and Sabah are generally mountainous and drained by an intricate system of rivers. Almost 70% of Sarawak consists of very steep areas. The highest peak is Mount Mulu with a height of 2371 m. The interior of Sabah has a series of mountain ranges and hills, the most prominent of which is the Crocker Range that rises abruptly to Mount Kinabalu with a height of 4175 m, and it is the highest mountain in South East Asia (Aminuddin et al., 1990). The mountainous area will affect several aspects of landuse in Malaysia such as erosion, degradation, siltation, and so on. It will also cause problem in the development of structure and infrastructure in particular areas (Aminuddin et al., 1990).

SOIL EROSION IN UPLAND AND CROP AREA

A large proportion of land described as upland and steep land whose soil erosion and the associated process of nutrient depletion are important forms of land degradation. The problem is exacerbated by failure to implement erosion control measures that are appropriate for the prevailing circumstances in a timely manner. Very often, cleared land is left exposed for extended periods before erosion control work is carried out. In some cases of highland areas, where high-value subtropical crops are produced, the effects of erosion are offset by the large application of manure and fertilizer. Although the practice of applying high rates of fertilizer and manure helps in maintaining reasonable levels of crop production, the unchecked losses of soil, nutrients, and chemicals through erosion processes contribute to downstream pollution and sedimentation. These are major environmental problems in some highland districts. Further improvements in erosion control, especially in terms of increased awareness, selection of appropriate measures and timely implementation, are necessary in view of the limited extent of good arable land. These measures as well as a total embrace of soil conservation will contribute in preserving the considerably large area of potentially productive upland.

DEVELOPMENT CONSTRAINTS

In general, farmers in steep land areas do not utilize good soil conservation practices, as they are usually motivated by short-term profit. As such in the Cameron Highlands catchment, soil loss is more than 125 kg/ha/year. This has caused extensive siltation of the hydroelectric dam downstream and shortened its life span to a mere third of initial projection. Moreover, the hydroelectric power generator cannot be operated during peak downpours as the sediment load is too high. The quality of drinking water is equally affected. Indiscriminate farming in such an area cannot be sustained. In some places, soil has to be brought from outside areas to replenish what was lost by erosion after a period of about 30 years. The removal of forest has also caused an increase of the surrounding temperature by 10°F or 2°C. However, in other steep areas, where cover crops are used, very low erosion rates are experienced. The practice of shifting cultivation on steep land was found to be ecologically stable and sustainable when the fallow period was about 10–15 years. Recently, however, the erosion problems have become severe as the fallow period decreased, in some cases to less than 3 years. Such a landuse is unproductive and wasteful, and a workable alternative farming system must be introduced (Aminuddin et al., 1990).

CONSTRUCTION PROBLEM IN STEEP LAND

A major constraint in steep areas is the slope length. This coupled with heavy tropical rainfall causes excessive runoff and erosion. Proper choice of crops on steep areas minimizes the developmental constraints. A technological package for growing rubber on such a slope is available. In contrast, such soils are not at all suitable for the cultivation of annual crops. Continued erosion under annual crops would reduce the soil depth and soil fertility as the organic matter and clays are removed. Other technological constraints include unfavorable effects of land clearing. In land development projects, areas as large as 2000 ha are cleared at one time mostly using heavy machines. Some research showed that mechanical clearing resulted in a lowering of CEC, organic carbon, and potassium (K) content in topsoil (Ling et al., 1979). CEC is a measure of the soil's ability to hold positively charged ions. It is a very important soil property influencing soil structure stability, nutrient availability, soil pH and the soil's reaction to fertilizers, and other ameliorants (Hazelton and Murphy, 2007). Besides that, mechanical clearing also compacted the soil and reduced infiltration, and removed vegetative cover, making it more vulnerable to erosion (Pimentel and Kounang, 1998). From an economic point of view, a major constraint is seen in the added cost of development, especially the cost of conservation measures. Moreover, as machinery usage is not encouraged in very steep areas, higher costs are incurred during manual land clearing and terracing (Aminuddin et al., 1990).

POORLY DEVELOPED INFRASTRUCTURE

Presently, the road system to the hinterland is poorly developed. Development of such infrastructure is extremely expensive because of dissected terrain. It is also difficult to justify when rural settlements are so widely scattered. As a result, farm inputs and basic human necessities are costly to deliver to these communities. Extension efforts to those are equally costly, arduous, and inefficient (Aminuddin et al., 1990).

NEGATIVE IMPACT TO THE ENVIRONMENT

Natural life systems or ecological systems are also among the most sensitive to environmental changes brought about by human activity. Such changes may directly impact species of plants and animals or indirectly through alteration to their habitat(s) and life support system. Besides that, it can also affect the areas identified as important or potentially important sources of groundwater supply. It should be avoided by those developments or activities, which have a high potential of contaminating groundwater or reduce its capacity for recharge. This is because once contaminated, groundwater is difficult to remediate. Similarly, turtle landings during the breeding season are affected by noise and light, which are likely to cause them to move elsewhere (Malaysia Ministry of Natural Resources and Environment, 2012).

WATER IN SOILS

The role of water content in soil material is a critical factor for its compaction to get maximum density. Currently, the conventional methods are unable to predict the proper moisture content for its compaction effort. For the strength parameter of soil, CBR mainly depends on moisture content. This physical property probably does not properly predict soil behavior under load conditions resulting from heavy traffic. Moisture content of soils is not properly investigated in design and construction stages. The dielectric constant of water is 78 (approximately) at 25°C at a microwave frequency of 100 MHz; this value varies with the amount of substrates present in soil materials. To get proper Proctor value of compaction effort, microwaves are able to provide proper moisture content of soil in terms of dielectric constant. It can provide valuable information as water is very sensitive to microwaves. Unsuitable soils are not desirable for construction; they need major modification and are costly.

LAND CHARACTERISTICS OF MALAYSIA

DAMAGES DUE TO SEISMIC HAZARD

Major earthquakes originating from these interplate boundaries (subduction zone) volcanic arcs have been felt in Malaysia. Sabah and Sarawak have experienced moderate earthquake of local origin that appeared to be related to several possible active faults. In 1976, in the district of Lahad Datu, an earthquake of magnitude 5.8 on the Richter scale developed some cracks in walls of buildings. Several flexible roads were also cracked as reported in the area. A four-storey police complex nearing completion suffered severe structural damages. An earthquake of magnitude 5.2 caused extensive damage to a four-storey teacher's quarter and was declared unfit for occupation. On May 2, 2004, a state near Miri in Sarawak similarly caused damage to the nonreinforced concrete buildings and cracks developed in the ground.

SOIL SETTLEMENT ISSUE

Bentong Lipis Road

Soil settlement is a major issue for Malaysian roads. Bentong Lipis Road was a project undertaken by Malaysian JKR (Jabatan Kerja Raya) in November 2001 and completed in October 2004. The subgrade soil was clayey in nature. The road was built on this clayey soil without proper treatment. Consolidation was taking place.

The maintenance cost was five million Malaysian Ringgit (average) per year (Unit, 2001). The original contract's cost was RM 140 million. The bottommost layer of pavement subgrade was not treated properly. The cost of repair each year was higher compared to the original construction cost of the project (Information from Maintenance unit, Road branch, Jabatan Kerja Raya [JKR/PWD] Head quarters).

UNSUITABLE SOIL MATERIALS

Unsuitable soil material exists in many parts of Malaysia, for example, Bukit Jalil, Kuala Lumpur, and the place of 16th Commonwealth games, held in the year 1998. Many structures were built in this area to facilitate the games. At one of the site support facilities, a crawler excavator moved on the soil surface to work. The ill-fated excavator sunk into the soil within a few minutes, as it was unable to move to a safer place. This depicts a clear picture of unsuitable soil materials in Malaysia. The surrounding areas of KL International Airport (KLIA) also have unsuitable soil material. During construction of roads in that locality (Road B20, Eastern access to KLIA), a replacement of unsuitable soil material by sand upto a depth of 3.0 m, was carried out. The depth was calculated by the Mackintosh Probe results.

SLOPE FAILURES

In Malaysia, usually many slope failures occur during the monsoon season after a pro-longed intense rainfall period. Recent cases showed that some major roads especially those constructed in rugged mountainous terrain were totally cut off during these slope failures. A typical example is the 112 km East–West highway linking the western and eastern region of Peninsular Malaysia (Lloyd et al., 2001). Failures of cut slopes along the highway are quite common because their natural formation is subjected to weathering and erosion. Surface water infiltration to the cut slope causes increased water pressure inside the soil and corresponding reduction in soil's shear strength.

CAVITIES AND FRACTURES

Subsurface cavities and fractures are associated with foundation and pilling problems. The topographical model was used to automatically extract minimum surface curvature, slopes, and pits. More than 14 regional faults affect the Kuala Lumpur limestone bedrock and the surface topography from South to North (Hashim and Islam, 2008). These fractures often show higher probability of piling and constructions problems.

SOILS AND CONSTRUCTION IN MALAYSIA

Peat soil is problematic soil in the construction field and is located all over the earth except in the arctic and desert area. The amount of this land is about 30 million ha, 5%–8% of the earth's total surface (Harden and Taylor, 1983). In South East Asia alone, it is about 23 million ha. Research was carried out to find ways of improving engineering techniques and using peat soil as a construction material by using different techniques and additives (Hebib and Farrell, 2003; Huat, 2006). A study on traditional fired bricks and stabilized compressed peat-based bricks both in experimental and field investigation construed that it will be possible to use peat soil as building wall materials. Such soils can help decrease material cost, create opportunities to use peat soil, reduce energy consumption, and minimize environmental damage.

During the rainy season, there is heavy rain in Malaysia, and in summer it is extremely hot. There is a definite need to investigate and develop the properties of compressed stabilized peat bricks to withstand extreme weather conditions, which are expected to intensify due to global warming.

Roads and Highways

The bottommost layer of a road is known as subgrade layer, which is constructed from natural soil or imported soil depending on the soil's properties. The investment in road industry is the largest in many countries. A well-constructed road lasts long and requires less maintenance, depending mainly on subgrade formation level (soil). Rutting deformation occurs at subgrade formation level, which is the base and bottommost layer of roads and highways for carrying design axle loads. All the deterioration of roads, especially rutting deformation occurs at subgrade level.

SOIL IMPROVEMENT AND STABILIZATION

Additives in Soils

Ordinary Portland Cement (OPC) is normally used as a soil additive particularly in the arid region, as it stabilizes clayey and sandy soils. In sediment soils, cement has the power to increase the plasticity index and to decrease the liquid intake, thus increasing the workability of soil. Theoretically, all kinds of soil can be stabilized with cement. But, with the increase of silt and clay content in the soil, more cement is required for stabilization of the soil. The addition of inorganic chemical stabilizers such as cement and lime has twofold effects on the soil, namely, acceleration of flocculation and promotion of chemical bonding. This bonding is subjected to characters of the additive (Deboucha et al., 2008) and strength of clay, and silt can develop up to 30-fold (Janz and Johansson, 2002). In the stabilizing process of soil, any kind of cement or lime can be used. OPC is commonly used (Janz and Johansson, 2002). In the case of soft and organic soil stabilization, adding additives such as lime, cement, and fly ash to soft soil helps develop its strength.

Ground Improvement for Infrastructure Development

Several soil improvement methods are available for the problematic soils in Malaysia (Raju and Yandamuri, 2010). Among them are jet grouting, deep soil mixing, vibro concrete column, and vibro stone column. Table 5.12 shows the soil problem in infrastructure development and the application of soil improvement.

TABLE 5.12
Soil Problem in Infrastructure Development and Corrective Techniques

Project Background	Problem/Soil Condition	Solution/Improvement
A project involved the construction of a 13 m diameter bored tunnel in Kuala Lumpur, over a distance of approximately 10 km. The tunnel will mainly function as a storm water storage and diversion channel as well as incorporating a 3 km triple deck motorway. Construction of a 13 m diameter bored tunnel over a distance of approximately 10 km. The cutter head of the tunnel boring machine (TBM) required maintenance at regular intervals. During such TBM stops (referred as *cutter-head intervention*), the slurry pressure is switched off and stability of the rock/soil face in front of the TBM relies on air pressure and the inherent strength of in situ rock/soil.	The geology encountered along the tunnel path was ex-mining soils and for limestone formation. Due to the existence of loose sandy material, there was a risk of ground disturbance and subsequent ground subsidence, if left untreated.	Jet grouting. The capping shield made of jet grout columns was designed to form a stable block at the cutter-head intervention location.
Three-storey commercial complex with two-level basement car park floors (about 7 m depth below the existing ground level) is under construction in the middle of Kuala Lumpur city centre. The project site is confined between a newly completed four-storey commercial lot, light rail transit track, and existing old warehouse.	Subsoil comprised loose silty sand deposits and ex-mining soils. Karstic limestone formation was found underlying these loose soil layers, with extremely varying rock-head levels ranging between 3 and 15 m below the existing ground level.	Deep soil mixing. The gravity wall is built to act as a temporary retaining structure during the basement excavation works. Wet deep soil mixing columns of 0.85 m diameter were interlocked at 0.75 m centers to form the rigid gravity wall block.
Huge sewage treatment plant is under construction in Penang Island, and after completion it will serve as a centralized sewage treatment facility. It will include 12 sequential batch reactor (SBR) tanks and associated process tanks.	The site was reclaimed from the sea and approximately half of the SBR tanks area was covered by former domestic landfill waste dumps (3–5 m thick).	Vibro concrete columns and deep soil mixing were utilized to support the SBR tanks on the ground.

(Continued)

TABLE 5.12 (*Continued*)
Soil Problem in Infrastructure Development and Corrective Techniques

Project Background	Problem/Soil Condition	Solution/Improvement
Modern expressway with dual three-lane carriageway forms the main interchange at Kampung Pasir Dalam to connect three distinct routes in the city. Due to site constraints at the interchange, high reinforced soil walls were constructed to form approaches and other ramps to the bridge according to the required design heights (maximum up to 13 m).	Subsoil conditions at Pantai Dalam interchange varied from very soft silts to soft sandy silts down to a depth between 5 and 12 m followed by hard sandy silts.	Vibro stone columns are used to support reinforced soil walls, which used to support the soil. The combination has proven economical and has intrinsic technical advantages; that is, the stone columns ensure relatively quick consolidation as the embankment is built, whereas the wall is constructed in stages (lifts) with wall panels placed progressively and adjusted for any movement.

Source: Raju, V. R., Yandamuri, H. K., *Proc. ICE-Ground Improv.,* 163, 251–263, 2010.

GROUND IMPROVEMENT OF SLOPE

Common techniques adopted for remedial work on failed fill slopes along the East–West highway included the reconstruction of slope for a stable gradient, improvement of overall drainage, reinforced slope, and retaining structures. In most cases, the remedial work involved a combination of the techniques (Anderson et al., 2000). Reconstruction of the failed fill slope consists of excavation and removal of debris and loose soil from the existing failure surface up to a stronger layer. The depth of excavation is determined from the Mackintosh probe results where 80 blows/300 mm is the limit. The new fill slope is then reconstructed in layers and compacted according to the standard Proctor compaction method.

1. In instances where seepage points are located within the fill slopes, sufficient discharge facility is provided in the stabilized fill area by installing layers of 300 mm thickness of sand drainage blanket. This sand layer is also incorporated between the excavated surface and the newly placed fill material.
2. Reinforced earth techniques are also employed for some remedial work for fill slopes at the East–West highway. This method applies to the combination of soil and reinforcing elements introduced into the body of the fill slope.

Fill slope remedial work, which involved realignment since cutting through the hill is inevitable, requires retaining structures in the form of bored piles. This method is applicable when there is a lack of space, and site conditions are restricted due to certain geological conditions. Installation of bored piles of 1000–1200 mm diameter spaced closely to retain a fill slope is quite common under the prevailing circumstances.

OTHER GROUND IMPROVEMENT

Other soil improvement methods associated with the soil problem are as follows:

1. Soil stabilization by applying modifiers
2. Surcharge
3. Removal of unsuitable soil by sand with geotextiles
4. Prefabricated vertical drain
5. Piling

SOIL INVESTIGATION

The basic reasons for conducting soil investigation are to determine the geotechnical engineering properties of the soil and to evaluate the groundwater level. It can be done by conducting in situ test and by collecting disturbed and undisturbed soil samples for testing in the laboratory. Also, soil investigation must be carried out in accordance with the specifications and guidelines stated in the approved standards.

Standard penetration test (SPT) and standpipe piezometer methods of field exploitation would be briefly discussed. SPT can be performed at an interval of 1.5 m by driving a split spoon of 50 mm diameter into the soil using a 65 kg hammer with a falling height of 760 mm. The numbers of blows for the initial 150 mm was recorded as seating blows, and the following numbers of blows for the next 300 mm penetration was recorded as the N value (or the blow count) of the soil strata encountered.

The standpipe piezometer consists of a tube with a porous filter element at the end that can be sealed into the ground at the appropriate level. It has a cylindrical (low air entry) porous element protected by performed rigid sheath about 3.5 mm diameter and 300 mm long. This element is connected to a 19 or 25 mm internal diameter pipe. The response time of this type of standpipe piezometer is comparatively low, but it generally does not become a significant factor until the soil permeability is less than 10–7 m/s. At this permeability, the response time should not be more than a few hours when the piezometer is installed within 150 mm diameter by 400 mm long sand pocket.

REFERENCES

Abdulah, F. and Rusea, G. (2009). Document of inherited knowledge on wild edible fungi from Malaysia. *Blumea* 54, 35–38.

Acosta-Martinez, V., Reicher, Z., Bischoff, M., and Turco, R.F. (1999). The role of tree mulch and nitrogen fertilizer on turfgrass soil quality. *Biol. Fertil. Soils* 29, 55–61.

Alam, S., Khalil, S., Ayub, N., and Rashid, M. (2002). In vitro solubilization of inorganic phosphate by phosphate solubilizing microorganism (PSM) from maize rhizosphere. *Int. J. Agric. Biol.* 4, 454–458.

Aminaton, M. and Fauziah, K. (2003). *Characterisation of Malaysian Residual Soils for Geotechnical and Construction Engineering*. Project Report. Universiti Teknologi Malaysia, Johor Bahru, Malaysia.

Aminuddin, B., Chow, W., and Ng, T. (1990). Resources and problems associated with sustainable development of upland areas in Malaysia. In: Blair, G. and R. Lefroy (Eds.) Technologies for Sustainable Agriculture on *Marginal Uplands in Southeast Asia*. Proceedings No. 33 Australian Centre for International Agricultural Research, Canberra, pp. 55–61.

Anderson, D., Youtcheff, J., and Zupanick, M. (2000). Asphalt binders. *Transportation in the New Millennium*, Transportation Research Board, Washington, DC.

Anderson, D.L., Kussow, W.R., and Corey, R.B. (1985). Phosphate rock dissolution in soil: Indications from plant growth studies. *Soil Sci. Soc. Am. J.* 49, 918–925.

Antunes, S.C., Pereira, R., Sousa, J.P., Santos, M.C., and Gon-calves, F. (2008). Spatial and temporal distribution of litter arthropods in different vegetation covers of Porto Santo Island (Madeira Archipelago, Portugal). *Eur. J. Soil. Biol.* 44, 45–56.

Arnold, A.E., Mamit, L.J., Gehring, C.A., Bidartondo, M.I., and Callahan, H. (2010). Interwoven branches of the plant and fungal trees of life. *New Phytol.* 185, 874–878.

Arpana, J. and Bagyaraj, D.J. (2007). Response of Kalmegh to an arbuscular mycorrhizal fungus and a plant growth promoting rhizomicroorganism at two levels of phosphorus fertilizer. *Am. Euras. J. Agric. Environ. Sci.* 2, 33–38.

Aubert, M., Hedde, M., Decaens, T., Bureau, F., Margerie, P., and Alard, D. (2003). Effect of tree canopy composition on earthworms and other macro-invertebrates in beech forests of Upper Normandy (France). *Pedobiologia* 47, 904–912.

Baldani, V.L.D., Baldani, J.I., and Döbereiner, J. (2000). Inoculation of rice plants with endophytic diazotrophs *Herbaspirillum seropedicae* and *Burkholderia* spp. *Biol. Fertil. Soil.* 30, 485–491.

Barrios, E. (2007). Soil Biota, ecosystem services and land productivity. *Ecological Economics.* 64, 269–285.

Bescansa, P., Imaz, M.J., Virto, I., Enrique, A., and Hoogmoed, W.B. 2006. Soil water retention as affected by tillage and residue management in semiarid Spain. *Soil Till. Res.* 87, 19–27.

Bird, S.B., Coulson, R.N., and Fisher, R.F. (2004). Change in soil and litter arthropod abundance following tree harvesting and site preparation in a loblolly pine (Pinus taeta L.) plantation. *Forest Ecol. Manag.* 202, 195–208.

Biswas, J.C., Ladha, J.K., and Dazzo, F.B. (2000). Rhizobial innoculation improves uptake and growth of lowland rice. *Am. J. Soil Sci. Soc.* 64(5), 1644–1650.

Bohluli, M., Teh, C.B.S., Husni, M.H.A., and Zaharah, A.R. (2014). Silt pit efficiency in conserving soil water as simulated by HYDRUS 2D model. *Pertanika J. Trop. Agric.* 37, 317–326.

Bongers, T. and Ferris, H. 1999. Nematode community structure as a bioindicator in environmental monitoring. *Trends Evol. Ecol.* 14, 224–228.

Brady, N.C. and Weil, R.R. (Eds.). (1999). *Nnature and Properties of Soils*. Prentice-Hall, Englewood Cliffs, NJ.

Brown, S. (1996). Soil mechanics in pavement engineering. *Geotechnique* 46(3), 383–426.

Burkill, I.H. (1966). *A Dictionary of the Economic Products of the Malay Peninsula*. Vol. I & II. Ministry of Agriculture, Kuala Lumpur, Malaysia.

Cakmakci, R., Donme, F., Aydin, A., and Sahin, F. (2006). Growth promotion of plants by plant growth promoting rhizobacteria under greenhouse and two different field soil conditions. *Soil Biol. Biochem.* 38, 1482–1487.

Chang, Y.S. (1997). Ethnomycology: A Malaysian perspective. In: *Ethnobiology, Proceedings for FORTROP'96 International Conference*, Vol. 3, November 25–28, 1996, Bangkok, Thailand, pp. 133–141.

Chang, Y.S. and Lee, S.S. (2001). Utilisation of wild mushrooms by the Temuans in Selangor, Malaysia. *Poster Presented at CFFPR 2001, 100 Year Celebration of Forestry Research*, October 1–3, 2001, Nikko Hotel, Kuala Lumpur, Malaysia.

Chin, F.H. (1981). Edible and poisonous fungi from the forests of Sarawak. Part I. *Sarawak Mus. J.* 29, 211–225.

Chin, F.H. (1988). Edible and poisonous fungi from the forests of Sarawak. Part II. *Sarawak Mus. J.* 60, 195–201.

Chung, H., Park, M., Madhaiyan, M., Seshadri, S., Song, J., Cho, H., and Sa, T. (2005). Isolation and characterization of phosphate solubilizing bacteria from the rhizosphere of crop plants of Korea. *Soil Biol. Biochem.* 37, 1970–1974.

Cocking, E.C. (2003). Endophytic colonization of plant roots by nitrogen-fixing bacteria. *Plant Soil.* 252, 169–175.

Crawley, M.J. (1989). Insect herbivores and plant population dynamics. *Ann. Rev. Enlomol.* 34, 531–64.

Daljit Singh, K.S., Arifin, A., Radziah, O., Shamsuddin, J., Hamid, A., Hazandy, Majid, N.A., Muhamad, N., Mohanaselvi, P., Halim, A., and Halizah, N. (2011). Assessing soil biological properties of natural and planted forests in the Malaysian tropical lowland dipterocarp forest. *Am. J. Appl. Sci.* 8(9), 854–859.

Deboucha, S., Hashim, R., and Alwi, A. (2008). Engineering properties of stabilized tropical peat soils. *EJGE.* 13, 1–9.

Dobbelaere, S., Vanderleyden, J., and Okon, Y. 2003. Plant growth promoting effects of diazotrophs in the rhizosphere. *Crit. Rev. Plant Sci.* 22, 107–149.

Dobereiner, J. (1993). History and new perspectives of diazotrophs in association with non legumes plants. *Symbiosis.* 13, 1–13.

Edwards, C.A. (1998). The use of earthworms in the breakdown and management of organic wastes. In: Edwards, C.A. (Ed.) *Earthworm Ecology.* CRC Press, Boca Raton, FL, pp. 327–354.

Elliot, L.F. and Lynch, J.M. (1995). The international workshop on establishment of microbial inocula in soils: Cooperative research project on biological resource management of the Organization for Economic Cooperation and Development (OECD). *Am. J. Alternative Agr.* 10, 50–73.

Ferris, H., Venette, R.C., and Lau, S.S. (1997). Population energetics of bacterial-feeding nematodes: Carbon and nitrogen budgets. *Soil Biol. Biochem.* 29, 1183–1194.

Fierer, N., Carney, K.M., Horner-Devine, M.C., and Megonigal, J.P. (2009). The biogeography of ammonia-oxidizing bacterial communities in soil. *Microb. Ecol.* 58, 435–445.

Franche, C., Lindstrom, K., and Elmerich, C. (2009) Nitrogen-fixing bacteria associated with leguminous and non-leguminous plants. *Plant Soil* 321, 35–59.

Garg, V.K., Kaushik, P., and Dilbaghi, N. (2006). Vermiconversion of waste water sludge from textile mill mixed with anaerobically digested biogas plant slurry employing *Eisenia foetida. Ecotoxicol. Environ. Saf.* 65, 412–419.

Gharibreza, M., Raj, J.K., Yusoff, I., Othman, Z., Tahir, W.Z.W.M., and Ashraf, M.A. (2013). Land use changes and soil redistribution estimation using 137 Cs in the tropical Bera Lake catchment, Malaysia. *Soil Till. Res.* 131, 1–10.

Goldstein, A.H. (1986). Bacterial solubilization of mineral phosphates: Historical perspective and future prospects. *Am. J. Altern. Agri.* 1, 51–57.

Greenland, D.J., Wild, A., and Adams, D. (1992). Organic matter dynamics in soils of the tropics—From myth to complex reality. In: Lal, R. and P.A. Sanchez (Eds.) *Myths and Science of Soils of the Tropic.* SSSA Special Publication no. 29. (pp. 17–33), Soil Science Society of America, and American Society of Agronomy, Madison, WI.

Guttiérrez-Zamora, M.L. and Martinez-Romero, E. (2001). Natural endophytic association between *Rhizobium etli* and maize (*Zea mays* L.). *J. Biotechnol.* 91, 177–126.

Hambali, H., Hafidi, M., Virolle, M.J., and Ouhdouch, Y. (2008). Growth promotion and protection against damping-off of wheat by two rock phosphate solubilizing actinomycetes in a P-deficient soil under greenhouse conditions. *Applied Soil Ecology* 40, 510–517.

Harden, J.W. and Taylor, E.M. (1983). A quantitative comparison of soil development in four climatic regimes. *Quatern. Res.* 20(3), 342–359.

Hashim, R. and Islam, S. (2008). Engineering properties of peat soils in peninsular, Malaysia. *J. Appl. Sci.* 8(22), 4215–4219.

Hazelton, P.A. and Murphy, B.W. (2007). *Interpreting Soil Test Results: What Do All the Numbers Mean?*. CSIRO publishing, Melbourne, Australia.

Hebib, S. and Farrell, E.R. (2003). Some experiences on the stabilization of Irish peats. *Can. Geotech. J.* 40(1), 107–120.

Hendrix, P.F., Crossley Jr., D.A., Blair, J.M., and Coleman, D.C. (1990). Soil biota as components of sustainable agroecosystems. In: Edwards, C.A., R. Lal, P. Madden, R.H. Miler, and G. House (Eds.) *Sustainable Agricultural Systems*. CRC press, SWCS, Ankeny, pp. 637–654.

Hernandez Martinez, F. and Al Tabbaa, A. (2009). Effectiveness of different binders in the stabilisation of organic soils.*Paper presented at the International Symposium on Soil Mixing and Admixture Stabilisat,* Okinawa, Japan.

Higa, T. (1995). Effective microorganisms: Their role in kyusei nature farming and sustainable agriculture. In: Parr, J.F., Hornick, S.B. and Simpson, M.E. (Eds.) *Proceedings of the Third International Conference on Kyusei Nature Farming.* U.S. Department of Agriculture, Washington, DC.

Hoyle, F.C., Baldock, J.A., and Murphy, D.V. (2011). Soil organic carbon—Role in rainfed farming systems: With particular reference to Australian conditions. In: R. Tow et al. (Eds.) *Rainfed Farming Systems*, Springer Science-Business Media BV, Amsterdam, The Netherlands.

Huat, B. (2002). Some mechanical properties of tropical peat and organic soils. *Paper presented at the Proceedings of the 2nd World Engineering Congress*, Sarawak, Malaysia.

Huat, B.B. (2006). Effect of cement admixtures on the engineering properties of tropical peat soils. *EJGE.* 11.

Huat, B.B., Maail, S., and Mohamed, T.A. (2005). Effect of chemical admixtures on the engineering properties of tropical peat soils. *Am. J. Appl. Sci.* 2(7), 1113.

Hunt, H.W., Coleman, D.C., Ingham, E.R., Ingham, R.E., Elliott, E.T., Moore, J.C., Rose, S.L., Reid, C.P.P., and Morley, C.R. (1987). The detrital food web in a short grass prairie. *Biol. Fertil. Soils* 3, 57–68.

Huntington, T.G. (2006). Evidence for intensification of the global water cycle: Review and synthesis. *J. Hydrol.* 319(1), 83–95.

Hurek, T., Reinhold-Hurek, B., Van Montagu, M., and Kellenberger, E. (1994). Root colonization and systemic spreading of *Azoarcus* sp. strain BH72 in grasses. *J. Bacteriol.* 176, 1913–1923.

Husni, M.H.A., Devi, S., Manas, A.R., Anuar, A.R., and Shamshuddin, J. (1995). Chemical variables affecting the lime requirement determination of tropical peat soils. *Commun. Soil Sci. Plant Anal.* 26(13 and 14), 2111–2122.

Hussin, A.S.M., Farouk, A., Ali, A.M., and Greiner, R. (2010). Production of phytate-degrading enzyme from Malaysian soil bacteria using rice bran containing media. *J. Agrobiotech.* 1, 17–28.

Ilstedt, U. (2002). Soil degradation and rehabilitation in humid tropical forests, Sabah, Malaysia. PhD thesis, Swedish University of Agricultural Sciences, Uppsala, Sweden.

Ingram, R.E., Trofymow, J.A., Ingram E.R. and Coleman, D.V. (1985). Interactions of bacteria, fungi and their nematode grazers: Effects on nutrient cycling and plant growth. *Ecological monograph.* 55(1), 199–140.

Ismail, B.S. and Shamshuddin, N. (2005). Effects of Alachlor and Metolachlor on microbial population in the soil. *Malays. J. Microbiol.* 1(1), 36–41.

Ismail, S.A. (2005). *The Earthworm Book*. Other India Press, Goa, p. 101.

Jais, H.M. and Hassan, H.M. (2008). Waste conversion to vermicast by *Eisenia foetida* given four types of organic substrates in the natural Malaysian environmental conditions. *J. Biosci.* 19(2), 63–72.

Janz, M. and Johansson, S. (2002). The function of different binding agents in deep stabilization. *Swedish Deep Stabilization Research Centre, Report No. 9*, Swedish Deep Stabilization Research Centre, Linkoping, Sweden, pp. 1–35.

Jeffrey, L.S.H., Sahilah, A.M., Son, R., and Tosiah, S. (2007). Isolation and screening of actinomycetes from Malaysian soil for their enzymatic and antimicrobial activities. *J. Trop. Agric. and Fd. Sci.* 35(1), 159–164.

Karlen, D.L., Wollenhaupt, N.C., Erbach, D.C., Berry, E.C., Swan, J.B., and Eash, N.S. (1994). Crop residue effects on soil quality following 10-years of no-till corn. *Soil Till. Res.* 31, 149–167.

Kennedy, I.R., Choudhury, A.T.M.A., and Kecskés, M.L., (2004). Non-symbiotic bacterial diazotrophs in crop-farming systems: Can their potential for plant growth promotion be better exploited? *Soil Biol. Biochem.* 36, 1229–1244.

Kim, K.Y., Jordan, D., and McDonald, G.A. (1998). *Enterobacter agglomerans*, Phosphate solubilizing bacteria and microbial activity in soil: Effect of carbon sources. *Soil. Biol. Biochem.* 30, 995–1003.

Kolenbrander, G.J. (1974). Efficiency of organic manure in increasing soil organic matter content. *Trans. Int. Congr. Soil Sci.* 2, 129–136.

Kundu, D.K. and Ladha, J.K. (1995). Enhancing soil nitrogen use and biological nitrogen fixation in wetland rice. *Exp. Agric.* 31, 261–277.

Lee, C.T., Ismail, M.N., Razali, F., Muhamad, I.I., Sarmidi, M.R., and Khamis, A.K. (2008) Application of effective microorganisms on soil and maize. *J. Chem. Nat. Resour. Eng.* 2, 1–13.

Ling, A., Tan, K., Tan, P., and Sofi, S. (1979). Preliminary observations on some possible post-clearing changes in soil properties. *Paper presented at the Seminar on Fertility and Management of Deforested Land*, Kota Kinabalu, Society of Agriculture, Sabah, Malaysia.

Lloyd, D., Anderson, M., Hussein, A., Jamaludin, A., and Wilkinson, P. (2001). Preventing landslides on roads and railways: A new risk-based approach. *Paper presented at the Proceedings of the ICE-Civil Engineering.* 144(3), 129–134.

Maene, L., Wan Sulaiman, W.H., Mohd. Mokhtaruddin, A.M., Maesschalck, G.G., and Lim, K.H. (1983). *Register of Soil Physical Properties of Malaysian Soils*. Technical Bulletin, Faculty of Agriculture. Universiti Pertanian Malaysia, Serdang, Malaysia.

Malaysia Ministry of Natural Resources and Environment (2012). *Guidelines for Siting and Zoning of Industry and Residential Areas*. Department of Environment Ministry of Natural Resources and Environment, Kuala Lumpur, Malaysia.

Mathieu, J., Rossi, J., Grimaldi, M., Mora, P.H., Lavelle, P., Rouland, C., and Rouland, A. (2004). Multi-scale study of soil macrofauna biodiversity in Amazonian pastures. *Biol. Fertil. Soils* 40, 300–305.

McBride, R. (2002). 2.9 Atterberg Limits. *Methods of Soil Analysis: Part 4 Physical Methods* (methodsofsoilan4), Soil Science Society of America, Fitchburg, WI, pp. 389–398.

Mirza, M.S., Rasul, G., Mehnaz, S., Ladha, J.K., So, R.B., Ali, S., and Malik, K.A. (2000). Beneficial effects of inoculated nitrogen-fixing bacteria on rice. In: Ladha, J.K. and P.M. Reddy (Eds.) *The Quest for Nitrogen Fixation in Rice*. International Rice Research Institute, Los Banos, Philippines, pp. 191–204.

Moraidi, A., Teh, C.B.S., Goh, K.J., Husni, M.H.A., and Fauziah, C.I. (2013). Soil organic C sequestration due to different oil palm residue mulches. In: Hamdan, J. and J. Shamshudin (Eds.) *Advances in Tropical Soil Science*. Vol. 2. Universiti Putra Malaysia, Serdang, Malaysia, pp. 169–186.

Moraidi, A., Teh, C.B.S., Goh, K.J., Husni, M.H.A., and Fauziah, C.I. (2015). Effect of four soil and water conservation practices on soil physical processes in a non-terraced oil palm plantation. *Soil Till. Res.* 145, 62–71.

Morgan, R.P.C. (2005). *Soil Erosion and Conservation*, 3rd ed. Blackwell Publishing, Oxford, UK.

Mutalib, A.A., Radziah, O, Shukor, Y., and Naher, U.A. (2012). Effect of nitrogen fertilizer on hydrolytic enzyme production, root colonization, N metabolism, leaf physiology and growth of rice inoculated with *Bacillus* sp. (Sb42). *Aust. J. Crop Sci.* 6(9), 1383–1389.

Naher, U.A., Radziah, O., Shamsuddin, Z.H., Halimi, M.S., and Mohd Razi, I. (2009). Isolation and characterization of indigenous diazotroph from rice plants grown in tanjong karang rice irrigation project. *Int J Agric Biol.* 11, 547–552.

Naher, U.A., Radziah, O., Shamsuddin, Z.H., Halimi, M.S., Razi, M.I., and Rahim, K.A. (2011). Effect of root exuded specific sugars on biological nitrogen fixation and growth promotion in rice (*Oryza sativa*). *Aust. J. Crop Sci.* 5(10), 1210–1217.

Naher, U.A., Radziah, O. and Panhwar, Q.A. (2013). Beneficial effects of mycorrhizal association for crop production in the tropics- a review. *Int. J. Agric. Biol.*, 15, 1021–1028.

Nahrul Hayawin, Z., Astimar, A.A., Anis, M., Hakimi Ibrahim, M., Abdul Khalil, H.P.S., and Inrahim, Z. (2012). Vermicomposting of empty fruit bunch addition of palm oil mill effluent solid. *J. Oil Palm Res.* 24, 1542–1549.

Noble, A.D. and Ruaysoongnern, S. (2010). The nature of sustainable agriculture. In: Dixon, R. and E. Tilston (Eds.) *Soil Microbiology and Sustainable Crop Production*. Springer Science and Business Media B.V., Berlin, Heidelberg, pp. 1–25.

Onweremadu, E.U., Asawalam, D.O., and Ibe, I.E. (2007). Changes in soil properties following application of composted sludge on an isohyperthermic kandiudult. *Res. J. Environ. Toxicol.* 1, 62–70.

Panhwar, Q.A., Radziah, O., Zaharah, A.R., Sariah, M., and Mohd Razi, I. (2012). Isolation and characterization of phosphorus solubilizing bacteria from aerobic rice. *African J. Biotechnol.* 11(11), 2711–2719.

Pankhurst, C.E. (1997). Biodiversity of soil organisms as an indicator of soil health. In: Pankhurst, C.E., B.M. Doube, and V.V.S.R. Gupta (Eds.) *Biological Indicators of Soil Health*. CAB International, Oxan, UK.

Paramananthan, S. (2012). *Keys to the Identification of Malaysian Soils Using Parent Materials* (Explanatory Notes), 2nd ed. Param Agricultural Soil Survey (M) Sdn. BHD, Selangor, Malaysia.

Parr, J.F. and Hornick, S.B. 1992a. Agricultural use of organic amendments: A historical perspective. *Amer. J. Alternative Agric.* 7, 181–189.

Pimentel, D. and Kounang, N. (1998). Ecology of soil erosion in ecosystems. *Ecosystems* 1(5), 416–426.

Plant & Soil Sciences eLibrary[PRO]. (2014). Soil genesis and development, lesson 6—Global soil resources and distribution. Accessed December 11, 2014. http://passel.unl.edu/pages/informationmodule.php?idinformationmodule=1130447033&topicorder=12&maxto=12&minto=1.

Pradhan, N. and Sukla, L.B. (2005). Solubilization of inorganic phosphates by fungi isolated from agricultural soil. *African J. Biotech.* 5(10), 850–854.

Provorov, N.A. and Tikhonovich, I.A. (2003). Genetic resources for improving nitrogen fixation in legume–rhizobia symbiosis. *Genet. Resour. Crop Evol.* 50, 89–99.

Provorov, N.A. and Vorobyov, N.I. (2010). Simulation of evolution implemented in the mutualistic symbioses towards enhancing their ecological efficiency, functional integrity and genotypic specificity. *Theor. Popul. Biol.* 78, 259–269.

Raju, V.R. and Yandamuri, H.K. (2010). Ground improvement for infrastructure projects in Malaysia. *Pro. ICE-Ground Improv.* 163(4), 251–263.

Reich, P.B., Oleksyn, J., Modrzynski, J., Mrozinski, P., Hob-bie, S.E., Eissenstat, D.M., Chorover, J., Chadwick, O.A., Hale, C.M., and Tjoelker, M.G. (2005). Linking litter calcium, earthworms and soil properties: A common garden test with 14 tree species. *Ecol. Lett.* 8, 811–818.

Reinhold, B. and Hurek, T., (1988). Localization of diazotrophs in the root interior with special attention to the kallar grass association. *Plant Soil* 110, 259–268.

Rogers, W.P. (1989). Nitrogenous components and their metabolism: Acanthocephala and Nematoda. In: Florkin, M. and B.T. Scheer (Eds.) *Chemical Zoology*. Vol. III. Academic Press, New York, pp. 379–428.

Sabrina, D.T., Hanafi, M.M., Nor Azwady, A.A., and Mahmud, T.M.M. (2009). Earthworm populations and cast properties in the soils of oil palm plantations. *MJSS*. 13, 29–42.

Saxton, K.E. and Rawls, W.J. (2006). Soil water characteristic estimates by texture and organic matter for hydrologic solutions. *Soil Sci. Soc. Am. J.* 70, 1569–1578.

Sayad, E., Hosseini, S.M., Hosseini, V., and Salehe-Shooshtari, M.H. (2012). Soil macrofauna in relation to soil and leaf litter properties in tree plantations. *J. Forest Sci.* 58(4), 170–180.

Sayed Abdul Rahman, S.A., Mohd Zain, S.N., Bilal Mat, M.Z., Sidam, A.K., Othman, R.Y., and Mohamed, Z. (2014). Population Distribution of Plant-parasitic Nematodes of Bananas in Peninsular Malaysia. *Sains Malaysiana* 43(2), 175–183.

Shainberg, I., Sumner, M.E., Miller, W.P., Farina, M.P.W., Pavan, M.A., and Fey, M.V. (1989). Use of gypsum on soils: A review. *Adv. Soil Sci.* 9, 1–111.

Shamsuddin, J. (1989). Lime requirements of highly weathered Malaysian soils. *Pertanika*. 12(1), 109–111.

Shamsuddin, J. (ed). (2006). *Acid Sulphate Soils in Malaysia*. UPM Press, Serdang, Malaysia.

Shamsuddin, J., Wan Noordin, W.D., Roslan, I., Fauziah, C.I., and Qurban, A.P. (eds.). (2015). *Ultisols and Oxisols: Enhancing Their Productivity for Oil Palm, Rubber and Cocoa Cultivation*. UPM Press, Serdang, Malaysia.

Shim, K.F. and Chua, Y.L. (1986). Studies on the protein requirement of the guppy Poeciliareticulate. *J. Aquaricult. Aquat. Sci.* 4, 79–84.

Siti Hajar, S. (2011). Bioactive microbial metabolites from Malaysian rainforest soil fungi as a source of new drugs candidates, MSc Thesis, Faculty of Pharmacy, Universiti Technologi Mara, Serdang, Malaysia.

Sultan, A.I. (1997). *Vermicology—The Biology of Earthworms*. Orient Longman Ltd, New Delhi, p 92.

Tahery, Y., Nor Aini, A.S., Hazandy, A.H., Abdullah, M.P., and Norlia, B. (2011). Status of root knot nematode on kenaf cultivated on Bris soil in Kuala Terengganu, Malaysia. *World Appl. Sci. J.* 15(9), 1287–2011.

Tan, K.Z., Radziah, O., Halimi, M.S., Khairuddin, A.R., Habib, S.H., and Shamsuddin, Z.H. (2014a). Isolation and characterization of rhizobio and plant growth-promoting rhizobacteria and their effects on growth of rice seedlings. *Am. J. Agricult. Biol. Sci.* 9(3), 342–360.

Tan, M.L., Ibrahim, A.L., Yusop, Z., Duan, Z., and Ling, L. (2014b). Impacts of land-use and climate variability on hydrological components in the Johor River basin, Malaysia. *Hydrol. Sci. J.* 60(5), 873–889.

Teh, C.B.S. (2012a). Aggregate stability of tropical soils in relation to their organic matter constituent and other soil properties. *Pertanika J. Trop. Agric.* 35, 135–148.

Teh, C.B.S. (2012b). The stability of individual macroaggregate size fractions of Ultisol and Oxisol soils. *J. Agricult. Sci. Technol.* 14, 459–466.

Teh, C.B.S. and Iba, J. (2010). Accuracy of the Saxton-Rawls method to estimate the soil water characteristics for minerals soils of Malaysia. *Pertanika J. Trop. Agric.* 33, 297–302.

Teh, C.B.S., Jamal, T., and Nuranina, S. (2005). Aggregate breakdown rates of some Malaysian soils and their relation to several aggregate properties. *MJSS.* 9, 1–13.

Teh, C.B.S. and Zauyah, S. (2001). The effects of empty fruit bunches on some soil physical properties after ten years of annual application. *Agro-Search Res. Bull.* 8, 27–30.

Thambirjah, J.J., Zulkifli, M.D., and Hashim, M.A. (1995). Microbiological and biochemical changes during composting of palm empty fruit bunches. Effect of nitrogen uplementation on the substrate. *Bioresource Technol.* 52, 133–144.

Thambirajah, J.J and Kuthubutheen, A.J. (1989). Composting of palm press fibre. *Biol. Wastes.* 27, 257–269.

Thamer, S.J. (2008). The effect of some plants extracts and essential oils on the workers of termites laboratory *microcerotermics gabriles* (Isoptera: Termitidae). *Bas. J. Vet. Res.* 7(2), 2008.

Tondoh, J.E., Monin, L.M., Tiho, S., and Csuzdi, C. (2007). Can earthworm be used as bio-indicators of land-use pertur-bations in semi-deciduous forest?. *Biol. Fertil. Soils.* 43, 585–592.

Trinick, M.J. (1980) Relationships amongst the fast growing rhizobia of *Lablab pur-pureus, Leucaena leucocephala, Mimosa* spp, *Acacia farnesiana* and *Sesbania grandiflora* and their affinities with other rhizobial groups. *J. Appl. Bacteriol.* 49, 39–53.

Tripathi, B.M., Kim, M., Singh, D., Lee-Cruz, L., Lai-Hoe, A., Ainuddin, A.N., Go, R., Rahim, R. A., Husni, M.H.A., Chun, J., and Adams, J.M. (2012). Tropical soil bacterial communities in Malaysia: pH dominates in the equatorial tropics tool. *Microb. Ecol.* 64, 474–484.

Tripathi, B.M., Lee-Cruz, L., Kim, M., Singh, D., Go, R., Shukor, N.A., Husni, M.H., Chun, J., and Adams, J.M. (2014). Spatial scaling effects on soil bacterial communities in Malaysian tropical forests. *Microb. Ecol.* 68(2), 247–258.

Tsukamoto, J. and Sabang, J. (2005). Soil macro-fauna in an *Acacia mangium* plantation in comparison to that in a primary mixed dipterocarp forest in the lowlands of Sarawak, Malaysia. *Pedobiologia* 49, 69–80.

Unit, E.P. (2001). *8th Malaysia Plan.* Economic Planning Unit, Kuala Lumpur, Malaysia.

Unkovich. (2003). David and Goliath: Symbiotic nitrogen fixation and fertilisers in Australian agriculture. In: *Proceedings of the 12th Australian nitrogen fixation conference.* Glenelg, SA, September 2003.

Vazquez, P., Holguin, G., Puente, M., Cortes, A.E., and Bashan, Y. (2000). Phosphate solu-bilizing microorganisms associated with the rhizosphere of mangroves in a semi arid coastal lagoon. *Biol. Fert. Soils.* 30, 460–468.

Watling, R. and Ginns, J. (1998). E.J.H. Corner, 1906–1996. *Mycologia* 90, 732–737.

Wei, Y.Y., Aziz, N.A.A., Shamsuddin, Z.H., Mustafa, M., Aziz, S.A., and Kuan, T.S. (2012). Enhancement of plant nutrient contents in rice straw vermicompost through the addition of rock phosphate. *Acta Biol. Malays.* 1(1), 41–45

Yahya, A. and Azawi. S.K.A. (1998). Occurrence of phosphate solubilizing bacteria in some Iranian soils. *Plant Soil.* 117, 135–141.

Young, I.M. and Crawford, J.W. (2004) Interactions and self-organisation in the soil-microbe complex. *Science* 304, 1634–1637.

Yusof, M.F., Abdullah, R., Azamathulla, H.M., Zakaria, N.A., and Ghani, A. (2011). Modified soil erodibility factor, K, for Peninsular Malaysia soil series. In: *3rd International Conference on Managing Rivers in the 21st Century: Sustainable Solutions for Global Crisi of Flooding, Pollution and Water Scarcity (RIVER 2011)*, December 6–9, 2011, Penang, Malaysia, pp. 799–808.

Zain, N.M.M., Rosli, B.M., Sijam, K., Morshed, M.M., and Awang, Y. (2013). Effects of selected herbicides on soil microbial populations in oil palm plantation of Malaysia: A microcosm experiment. *African J. Microbiol. Res.* 7(5), 367–374.

Zainab, H. (1977). Complexing Ability of Soil Organic Matter with Mineral Elements and Interactions with Nutrients, M.Sc. Agric. Thesis. State University of Ghent, Belgium.

Zakaria, Z., Mohamed, A.R., Mohd Salih, N.H., and Abu Mansoor, S.N. *(2013)*. Total nitrogen content from earthworm (Eisenia foetide) using the Kjeldahl method. *IIUM Eng. J.* 14(1), 43–51.

Zou, X. and Bashkin, M. (1998). Soil carbon accretion and earthworm recovery following revegetation in abandoned sugarcane fields. *Soil Biol. Biochem.* 30, 825–830.

6 Agricultural Soil Contamination

B.M. Firuza and M.B. Yusuf

CONTENTS

MEANING OF AGRICULTURAL SOIL CONTAMINATION

INTRODUCTION

The outermost layer of the earth's crust is extremely thin when compared with the rest of the crust. It is hardly half a meter thick; yet, human existence depends primarily on it. Within this very thin layer is the soil (Banwart, 2011). Soil is not just dirt but it is a mixture of minerals, air, water, and organic materials, such as roots, decaying plant parts, fungi, earthworms, bacteria, and microorganisms (Doran and Parkin, 1994; Ellison, 2006). An acre of healthy soil can contain 900 lb of earthworms, 2400 lb of fungi, 1500 lb of bacteria, 133 lb of protozoa, 8900 lb of arthropods and algae, and in some cases, small mammals. Soil offers shelter and habitat for numerous organisms and is the basis for agriculture (Hillel, 2008; Koch et al., 2013). Soils are formed by solid resources recycling and complex processes of the solid crust modifications (Brady and Weil, 2002). Thus, soil properties may vary from region to region depending on the prevailing climatic, bedrock composition, and anthropogenic factors (Kelly and Yonker, 2005).

In some cases, the amounts of some chemicals, nutrients, or elements in the soils may exceed the soil threshold levels for optimum soil productivity. When such a level is found to be significantly higher than that of the *normal level* (background level), and when it is likely to be a risk to people's health and the soil, such a soil is considered to be contaminated (Food and Agricultural Organization, 2007; Feller et al., 2012). Therefore, the contamination of agricultural soils with harmful substances that can adversely affect the quality of the soil and the health of farmers and/or those working on it is referred to as agricultural soil contamination (Environmental Protection Agency, 2011). Soil contaminants are either solid or liquid hazardous substances mixed with the naturally occurring soil (León et al., 2009). Usually, contaminants in the soil are physically or chemically attached to soil particles, or they are trapped in small spaces between the soil particles. They occur as a result of both natural and human action.

The naturally occurring processes (weathering of rocks or geothermal activity) of the concentration of hazardous substances such as arsenic, mercury, and lead in the agricultural soil are so slow and at times imperceptible (Science Learning Hub, 2007–2014). Such substances become toxic if they are at high concentrations in the soil. Thus, many minerals are needed by organisms to be able to exist, but they are often needed in small quantity (Roetter et al., 2007). Other possible harmful substances in the soils including chemical fertilizer, pesticides, and other amendments emanate from improper agricultural practices (Bastida et al., 2008; Ezcurra et al., 2011). The most common contaminants that deteriorate soil quality and make it incapable for later use are harmful pesticides and insecticides. For example, the use of organochlorine DDT, which was widely used as a pesticide but now banned. Similarly, in the past, the use, handling, and storage of hazardous chemicals were not as controlled as they are today (European Commission, 2013). People in those days were not so aware of the damage being done to the soil or the effects these chemicals could have on their health (Environmental Protection Agency, 2011). Besides, the use of pesticides in agricultural practices holds chemicals in the soils for a long time. Such chemicals might affect the beneficial organisms in the soils such as earthworms and consequently lead to poor soil quality (Bastida et al., 2008; Ezcurra et al., 2011). Therefore, the concern over agricultural soil contamination stems primarily from health risks, direct contact with the contaminated soil, and from secondary contamination of water supplies within and underlying the soil (Environmental Protection Agency, 2011).

TYPES OF AGRICULTURAL CONTAMINANTS

Agricultural soil contaminants typically arise from natural and human activities. Weathering, application of pesticides and herbicides, leaching of waste from landfills, or direct discharge of industrial waste into agricultural sites are examples.

However, the type of contaminants in agricultural soil is linked with biological and geochemical cycles and is influenced by human activities such as agricultural practices, industrial activities, and waste disposal. In Malaysia, agricultural soil contamination are often linked to waste from oil palm industries, whereas heavy metals in soils are inherited from the parent materials or are added through the use of organic and chemical fertilizer and pesticide (Zauyah et al., 2004; Ismail et al., 2011).

Factors That Determine a Number of Contaminants in the Soils

The amount of contaminants in the soils that may have direct effects on plants, and animals, as a result of their uptake by their bodies is referred to as bioavailability portion (Shayler et al., 2009). However, some of the contaminants found in soils are not biologically available. Moreover, the bioavailability of contaminants in the soils depends on prevailing soil and site characteristics. For instance, site prevailing conditions determine the degree to which contaminants are tightly held by soil particles and their solubility (i.e., ability to dissolve in water). Greater solubility typically suggests that the greater proportion of the contaminants is bioavailability and therefore is much likely to leach out of the soils. Similarly, some chemicals can have *aging effects*, that is, the longer they remain in soils, the less the bioavailability.

What Happens to Contaminant in the Soils

Once contaminants are present in soils, some can undergo chemical changes such as organic (carbon-based), and others may be degraded into products with more or less toxics than their original compound. Several many others such as metals are hardly broken up, but their characteristics may change into a medium that can be easily absorbed by plants and animals (Shayler et al., 2009). As a whole, the availability of contaminants in the soils and their ability to be absorbed by plants and animals are determined by the characteristics of the soil. Accordingly, different contaminants vary in their propensity to evaporate (volatilize) into the air and/or bind tightly to the soils particles. Some significant characteristics of soils that affect the behaviors of soil contaminants are the soil texture (in the form of its mineralogical and clay contents), chemical (in the form of the soil pH), the amount of organic content in the soil, soil's temperature, moisture, and the presence of other soil chemical properties (Bronick and Lal, 2005; Araújo et al., 2013).

The Distribution of Contaminants in the Soils

The distribution of contaminants in soils released due to human agricultural practices is correlated with the site and soil constituents. However, movement of soil contaminants within an agricultural soil is dependent upon the movement of air and water. For instance, wind may carry chemicals and may deposit them on new surface soils, whereas such deposited chemicals can be mixed when tilling the soils (De Alba et al., 2004). Similarly, the movement of surface water also influences how soil contaminants spread from their sources (Hepperle and Schulin, 2008).

Accordingly, many pesticides and amendments used for improving soil fertility in agricultural farms could be found in residential soils. This often occurs due to the conversion of agricultural lands for residential purposes. Similarly, deposition of chemicals, such as arsenic and lead, by aerial techniques on agricultural fields may result in soil contamination in residential areas (Hartemink, 2006).

How Farms Get Contaminated

The farm can become contaminated when hazardous substances are leaked, spilled, or disposed off. Often, the contamination was unintentional or occurred despite following recommended management practices.

How Plants Get Contaminated

There are basically three ways through which agricultural plants (crops) could get contaminated (Shayler et al., 2009). The significance of this information is to help farmers select the best crop types for a particular situation, particularly garden crops.

1. *Deposition by air*: Airborne contaminants end up mostly on plants. This is true especially of leafy crops, which have high surface areas in contact with airborne particles.
2. *Uptake through plant roots*: In most crops, particularly in garden crops, unless the soil is acidic, (low pH) or very low in organic matter, lesser amount of lead substance is taken by garden crops through their roots from contaminated soils. However, roots are likely to have a higher concentration of lead substances than the plant stems, leaves, and fruits (seeds). However, the plant fruits had the lowest amount of lead substance than that of all plant parts. Some heavy metals of serious health concern, including cadmium substances, are more readily taken up into plant roots and tops from contaminated soils.
3. *Direct contamination by agricultural soil*: Root and tuber crops are more likely to be contaminated than other types of crops because they are in direct contact with soil. Leafy vegetables (lettuce, spinach, and collard greens) are also easily contaminated by soils splash and dust. Washing leafy crops can remove up to 80% of lead contamination, and much of the lead can be removed from vegetables such as carrots and potatoes by peeling. However, in a situation where lead contamination is moderate to severe, growing these types of crops directly in the contaminated soil is probably not the best choice.

There are several natural barriers that can limit heavy metal transfer into crops:

1. *Soil–root barrier*: Some toxic metals (such as lead) have low solubility in most soils and do not readily enter the plants through roots.
2. *Root–shoot barrier*: Most toxic metals bind relatively stronger in roots, and movement to other plant parts is limited.

3. *Shoot–fruit barrier*: Most toxic metals are largely excluded from entering the reproductive parts (fruit, seeds) of a crop, remaining instead in the vegetative parts.

How Farm Families Are Exposed to Soil Contaminants

Farm families may be exposed to soil contaminants through the processes of eating and/or drinking (ingestion), skin contact (dermal exposure), and breathing (inhalation). However, the processes of human exposure to soil contaminants vary with contaminant types and soil conditions, as well as with human activities at the particular site (Environmental Protection Agency, 2011; European Commission, 2013)

1. *Ingestion (eating or drinking)*: Most people, particularly children, accidentally ingest small amounts of soils in their daily activities especially those working in farms, gardens, or those playing on agricultural fields. Children generally ingest more soils than the adults, as a result of their frequent land-to-mouth behavior (Environmental Protection Agency, 2011). Accidental ingestion may also occur in adults (e.g., by eating vegetables with some soil still attached). In some part of the world agricultural regions, adults also deliberately eat soil (agricultural land soils) for a number of cultural reasons. Similarly, animals raised for food might take in contaminants from soils. Therefore, eating such animals' products such as its meat, eggs, and milk may expose people to such contaminants. Drinking water may contain soil contaminants that were directly discharged into the water sources or entered the water sources through surface runoff and/or indirectly leached from the soils into groundwater. Thus, direct ingestion is the most important partway for human exposure to agricultural soil contaminants, although other specific partway have some importance in certain situations.
2. *Inhalation (breathing)*: Farm families may also inhale soil contaminants that are bounded to soil particles by working with soils and become airborne, for example, windblown dust (soil contaminants that evaporate from the soils) (Environmental Protection Agency, 2011). Very small particles may lodge in the lungs, and there is a chance that contaminants may be absorbed into the bloodstream. Compared to ingestion, this is far less significant source of exposure but may be relevant to those exposed repeatedly over a long period of time.
3. *Skin contact (dermal absorption, cutaneous, or transcutaneous absorption)*: Some soil contaminants particularly pesticides can pass through human skin and enter the body, or it may become part of the air that people breathe through vaporization from the underlying groundwater.

CAUSES OF AGRICULTURAL CONTAMINANTS

Agricultural soil contamination comes about with the introduction of enormous quantities of xenobiotic (human-made) chemicals into the soil or with alteration in the natural soil environment as a result of man activities (Stockmanna et al., 2013).

It is typically caused by both natural and human activities such as poor farm management and/or inappropriate application of agricultural chemicals (Connor and Mínguez, 2012).

The Malaysian agriculture is characterized by two distinct subsectors, namely the plantation subsector and smallholders subsector. The major crops grown are oil palm, rubber, rice, mixed horticulture, coconut, and orchard. The overdependence of these primary commodities has made it necessary for the importation and usage of large quantities of manufactured chemical fertilizer in order to sustain crop production. The most common chemicals involved are pesticides, lead, and other heavy metals. However, owing to the divergent nature of soil contaminants, site and soil conditions, the significant impact of soil contaminants depend on the site-specific conditions of a particular farm (Zingore et al., 2007; Maimon et al., 2009). Some of the possible causes include the following.

USE OF PESTICIDES

Use of pesticides, fertilizers, and insecticides in agricultural land uses. Pesticides are chemicals used as insecticides, fungicides, herbicides, and poisons for rodents. These products are made of chemicals that are not originally found in nature and hence lead to soil contamination. Use of pesticides in agriculture retains chemicals in the environment for a long time. These chemicals also affect beneficial organisms such as earthworms in the soil and lead to poor soil quality.

FERTILIZERS USAGE

The use of organic and inorganic fertilizers might increase the level of some soil contaminants in the soils. Phosphate fertilizer, for instance, is made from phosphate rock that is rich in cadmium, whereas organic fertilizer contains a relatively high proportion of zinc and/or copper.

HIGH TRAFFIC AREAS

A farm's distance from roadways and traffic influences the amounts of certain chemicals in the soil, chiefly lead. For example, metal-contaminated dust on roads can wash into the soil as part of rain-induced runoff. The Food and Fertilizer Technology Center warns that heavy metal contamination cannot only reduce crop yields due to poor soil quality but also result in the crops absorbing the metals. Similarly, studies of the soils around Perlis, Malaysia, to assess heavy metal contamination distribution due to industrialization, urbanization, and agricultural activities, indicate that industrial activities and traffic emission represent the most important sources of contaminant (Ripin et al., 2014).

LANDFILLS/GARBAGE DUMPS

Some soil contaminants such as pesticides, solvents, lead, and petroleum products, including heavy metals, are leached out from landfills and garbage disposal sites.

However, a number of chemicals in agricultural fields located close to waste disposal sites (currently or in the past) may depend on the specific site conditions of a farm and the types of materials disposed of at that site.

INDUSTRIAL WASTE PRODUCTS

A substantial percentage of agricultural soil contamination is caused by industrial waste products. For instance, improper disposal of waste (on or near agricultural sites) contaminates the soil with harmful chemicals; such contaminants can affect plants, living organism, and local water supplies. Toxic fumes from the regulated landfills contain chemicals that can fall back to the earth (agricultural farm) in the form of acid rain and damage the soil profile.

ACID RAIN

When pollutants present in the air mix up with the rain and fall back on the ground, the contaminated water could dissolve away some of the important nutrients found in the soil and could change the structure of the soil. It can also harm living organisms in the soils and humans who come into contact with it especially by touching. Also include wrong or inappropriate soil management systems and poor irrigation practices.

EFFECTS OF AGRICULTURAL CONTAMINANTS

Chemical utilization has gone up tremendously because technology has started providing us with modern pesticides and fertilizers (Environmental Protection Agency, 2011). Agricultural soil contamination occurs where intensive agricultural activities have introduced excessive amounts of contaminants. Soils have only a limited ability to process these contaminants through filtering or transformation. Once this ability is exceeded, issues such as soil and water contamination, human contact with contaminated soil, plants taking up contaminants, and danger on human health become more significant.

EFFECTS ON AGRICULTURAL LAND

Chemical utilization has gone up tremendously because technology has started providing us with modern pesticides and fertilizers (Srinivasarao et al., 2014). Agricultural soils are, therefore, full of chemicals that are not produced in nature and cannot be broken down by it (E-SchoolToday, 2010). Chemicals can damage the composition of the soil and can make it easier to erode by water and air (Sparks, 2005). Agricultural processes such as the application of fertilizers increase crop yields and soil contaminants that impact the soil quality. Pesticides also harm plants and animals by contaminating the soil. When these chemicals seep into the ground after mixing with water, they slowly reduce the fertility of the soil. Runoff of these chemicals by rainwater and/or irrigation water can also contaminate the local water sources.

The chemicals present in the soil due to contamination are toxic and can decrease the natural fertility of the soil, thereby decreasing the soil yield and making it unfit for cultivation. Similarly, agriculture on contaminated soil produces fruits and vegetables that lack quality nutrients, and consumption of these may be poisonous and can cause serious health problems to people consuming them.

Effects on Living Organism in the Soils

Soil contamination may lead to the death of some living organisms found in the soils, such as earthworms, which may lead to alteration in the structure of the soils (Shayler et al., 2009; Environmental Protection Agency, 2011). It can also force other predators to move to other places in search of food; in other words, this can affect the food chain and can create additional problems of soil erosion.

Effects on Plants (Crops)

The balance in soil quality is affected due to soil contamination. Plants mostly are unable to adapt to changes in the chemistry of the soil in short-time period (Zhao et al., 2007). This poses a significant risk to the health of plants (crops).

For example, animal health, plant vigor, microbial processes, and the overall soil health are directly dependent on the amount of soil contaminants present in agricultural lands, which may impact negatively in extreme circumstances. Change in plants' metabolic processes and reduced yields and damage to crops are caused by the presence of high-concentration soil contaminants (Shayler et al., 2009). Similarly, a relatively low concentration of some soil contaminants may change the soil chemistry and impact negatively on plants and organisms that depend on the soil for their nutrition and habitat.

However, it is important to note that such consequences of soil contaminants on the animals, plants, and microbes, including soil, in any agricultural region are a function of many factors including the soil properties, the intensity of contamination, and the types of contaminants, as well as the sensitivity and resilience nature of soil organism to the existing contamination (European Commission, 2013). For instance, legume plants (bean, peas, lentils, peanuts, etc.) are nitrogen-fixing crops. Nitrogen is an important soil nutrient required for optimum plant growth and production. Legume plants fixed the soil nitrogen through a process of symbiotic relationship with *Rhizobium* bacteria in their root nodules. The bacteria are sensitive to zinc contaminant because it can disrupt the nitrogen fixation process.

Possible Health Effects of Exposure

Besides its impact on crop health, agricultural soil contamination has major consequences on human health (Environmental Protection Agency, 2011). Consumption of crops cultivated on contaminated soil causes health hazards. This could be explained by the occurrence of small and terminal illness. Long-term exposure to contaminated soil affects the genetic makeup of the body and causes many congenital illnesses and chronic health diseases (European Commission, 2013). Chronic exposure to agricultural chemicals can be carcinogenic (Shayler et al., 2009). Mercury causes higher incidences of kidney damage. Organophosphates can lead to a chain of responses leading to neuromuscular blockage. Chlorinated solvents induce damages to liver and kidney, and induce depression of the central nervous system (Environmental Protection Agency, 2011).

It is significantly important to note that for any exposure to agricultural soil contaminants, the probability of the contaminant to cause health effects depends on its toxicity (i.e., how harmful the contaminant is to humans), the amount of the contaminant that is in contact, and the duration of exposure (Science Daily, 2014). Minor hypothetical dynamics comprise the health status of a person, his diet, age, gender, lifestyle, family traits, and so on.

Thus, the way people respond due to their exposure to agricultural soil contaminant is a function of the aforementioned factors. As the whole, young children are more vulnerable compared to adults, probably, because they ingest and absorb more of the ingested contaminants (E-SchoolToday, 2010). Moreover, children eat, drink, and breathe more in relation to their body sizes than adults, besides, their bodies are still developing and more vulnerable to soil contaminants (European Commission, 2013).

METHODS OF CONTROLLING AGRICULTURAL SOIL CONTAMINATION

Contamination control is a generic term for all activities aiming to control the existence, growth, and proliferation of contamination in agricultural farms (E-SchoolToday, 2010). The aim of agricultural soil contamination control activities is to permanently ensure a sufficient level of cleanliness in the agricultural soil environment. Maintaining, reducing, or eradicating viable and nonviable agricultural soil contamination in order to maintain a sustainable soil quality for maximum and healthy crop yields accomplish this.

Depending on the extent and volume of agricultural soil contamination, it can be remedied by reducing the use of chemical fertilizer and pesticides. An organic method of farming, which does not use chemical-laden pesticides and fertilizers, are to be supported. Cultivation of plants that can remove the contaminants from the soil particularly in gardening should be encouraged.

The best way to solve problem is to understand it. The greatest step forward is to learn about it. Therefore, contamination awareness programs should be undertaken for farmers to educate and to make them aware of the harmful effects of inappropriate farm management for sustainable soil and crop yields.

Agricultural soil contamination can also be controlled by avoiding deforestation and promoting forestation, suitable and safe disposal techniques of wastes including industrial wastes, recycling paper, plastics, and other materials, and ban on the use of plastic bags, which are a major cause of contamination. However, the road ahead is quite long, and the prevention of soil contamination will take many more years.

RECOMMENDATIONS

Since there is no clear line of what is considered *safe* or threshold limit for optimum soil health, to be referred too if contaminant amount exceed or are lower than. It is recommended that effective management measures that aim at reducing soil contaminants should be employed. This imperative, and issue of serious health concern especially, in agricultural areas where children mostly play and farms where fruits or vegetables are grown for food. A legal framework to identify the extent of contaminated soils and to deal with this environmental problem should be encouraged, especially in developing countries where some of them are undergoing significant industrialization.

REFERENCES

Araújo, A. S. F., Cesarz, S., Leite, L. F. C., Borges, C. D., Tsai, S. M., and Eisenhauer, N. (2013). Soil microbial properties and temporal stability in degraded and restored lands of Northeast Brazil. *Soil Biology and Biochemistry*, 66(0), 175–181. doi: 10.1016/j. soilbio.2013.07.013.

Banwart, S. (2011). Save our soils. *Nature*, 474, 151–152. doi: 10.1038/474151a.

Bastida, F., Zsolnay, A., Hernández, T., and García, C. (2008). Past, present and future of soil quality indices: A biological perspective. *Geoderma*, 147(3–4), 159–171. doi: 10.1016/j. geoderma.2008.08.007.

Brady, N. C., and Weil, R. R. (2002). *The Nature and Properties of Soils*. Thirteenth edition, Upper Saddle River, NJ: Pearson Education.

Bronick, C. J., and Lal, R. (2005). Soil structure and management: A review. *Geoderma*, 124(1–2), 3–22. doi: 10.1016/j.geoderma.2004.03.005.

Connor, D. J., and Mínguez, M. I. (2012). Evolution not revolution of farming systems will best feed and green the world. *Global Food Security*, 1(2), 106–113. doi: 10.1016/j. gfs.2012.10.004.

Doran, J. W., and Parkin, T. B. (1994). Defining and assessind soil quality. In J. W. Doran and T. B. Parkin (Eds.), *Defining soil quality for a Sustainable Environment* (Vol. 35). Madison, WI: Soil Science Society of America Special Publication.

De Alba, S., Lindstrom, M., Schumacher, T. E., and Malo, D. D. (2004). Soil landscape evolution due to soil redistribution by tillage: A new conceptual model of soil catena evolution in agricultural landscapes. *CATENA*, 58(1), 77–100. doi: 10.1016/j. catena.2003.12.004.

Ellison, K. (2006). The nature of farms. *Frontiers in Ecology and the Environment*, 4(5), 280–282. doi: 10.1890/1540-9295(2006)004[0280:TNOF]2.0.CO;2.

Environmental Protection Agency. (2011). Soil contamination | Superfund | US EPA. Retrieved from: http://www.epa.gov/superfund/students/wastsite/soilspil.htm (accessed June 6, 2016).

E-SchoolToday. (2010). What is soil contamination. Retrieved from: http://eschooltoday.com/ pollution/land-pollution/what-is-soil-contamination.html (accessed June 6, 2016).

European Commission. (2013). IN-DEPTH REPORT soil contamination: Impacts on. *Human Health. Environment. Science for Environment Policy*. Retrieved from: http://ec.europa. eu/environment/integrat...earch/newsalert/pdf/IR5.pdf (accessed June 6, 2016).

Ezcurra, R., Iráizoz, B., Pascual, P., and Rapún, M. (2011). Agricultural productivity in the European regions: Trends and explanatory factors. *European Urban and Regional Studies*, 18(2), 113–135. doi: 10.1177/0969776410381037.

Feller, C., Blanchart, E., Bernoux, M., Lal, R., and Manlay, R. (2012). Soil fertility concepts over the past two centuries: The importance attributed to soil organic matter in developed and developing countries. *Archives of Agronomy and Soil Science*, 58, 3–21.

Food and Agricultural Organization. (2007). World reference base on soil resources. *World Soil resources report* (Vol. 20) Roman, Italy: Food and Agricultural Organization of the United Nation.

Hartemink, A. E. (2006). Assessing soil fertility decline in the tropics using soil chemical data. *Advances in Agronomy*, 89, 179–225. doi: 10.1016/S0065-2113(05)89004-2.

Hepperle, E., and Schulin, R. (2008). Soil monitoring and evaluation in the interest of land use planning. *Landscape and Urban Planning*, 88(2–4), 45. doi: 10.1016/j.landurbplan.2008.09.001.

Hillel, D. (2008). Soil in the environment crucible of terrestrial life. Retrieved from http://www.sciencedirect.com/science/book/9780123485366 (accessed June 6, 2016).

Ismail, B. S., Sameni, M., and Halimah, M. (2011). Evaluation of herbicide pollution in the kerian ricefields of perak, Malaysia. *World Applied Sciences Journal*, 15(05–13).

Kelly, E. F., and Yonker, C. M. (2005). Factors of soil formation|time. In D. Hillel (Ed.), *Encyclopedia of Soils in the Environment* (pp. 536–539). Oxford: Elsevier.

Koch, A., Mcbratney, A., Adams, M., Field, D., Hill, R., Crawford, J., Minasny, B. et al. (2013). Soil security: Solving the global soil crisis. *Global Policy (Survey Article)*, 1–8. doi: 10.1111/1758-5899.12096.

León, V., Fraschina, J., and Busch, M. (2009). Rodent control at different spatial scales on poultry farms in the province of Buenos Aires, Argentina. *International Biodeterioration & Biodegradation*, 63(8), 1113–1118. doi: 10.1016/j.ibiod.2009.08.004.

Maimon, A., Khairiah, J., Ahmad Mahir, R., Aminah, A., and Ismail, B. S. (2009). Comparative accumulation of heavy metals in selected vegetables, their availability and correlation in lithogenic and nonlithogenic fractions of soils from some agricultural areas in Malaysia. *Advances in Environmental Biology*, 3(3), 314–321.

Ripin, S. N. M., Hasan, S., Lias, M., Kamal, N., and Hashim, M. (2014). Analysis and pollution assessment of heavy metal in soil, Perlis. *The Malaysian Journal of Analytical Sciences*, 18(1), 155–161.

Roetter, R., Keulen, H., Kuiper, M., and Verhagen, J. (2007). Agriculture in a dynamic world. In R. Roetter, H. Keulen, M. Kuiper, J. Verhagen and H. H. Laar (Eds.), *Science for Agriculture and Rural Development in Low-income Countries* (pp. 1–6), Dordrecht, Netherlands: Springer.

Science Daily. (2014). Soil contamination. Retrieved from: http://www.sciencedaily.com/articles/s/s...es/s/soil_contamination.html (accessed June 6, 2016).

Science Learning Hub. (2007–2014). Soil, farming and science, faculty of education The University of Waikato—Te whare wananga o waikato private bag 3105 Hamilton 3240 New Zealand. Retrieved from: http://sciencelearn.org.nz/user/activate/4ec2152d10fb78 44cb1bcfafa363979b/20797 (accessed June 6, 2016).

Shayler, H., McBride, M., and Harrison, E. (2009). Sources and impacts of contaminants in soils—Cornell waste management institute. Retrieved from: http://cwmi.css.cornell.edu/sourcesandimpacts.pdf (accessed June 6, 2016).

Sparks, D. L. (2005). Toxic metals in the environment: The role of surfaces. *Elements*, 1(4), 193–197. doi: 10.2113/gselements.1.4.193.

Srinivasarao, C. H., Venkateswarlu, B., Lal, R., Singh, A. K., Kundu, S., Vittal, K. P. R., and Patel, M. M. (2014). Long-term manuring and fertilizer effects on depletion of soil organic carbon stocks under pearl millet-cluster bean-castor rotation in Western India. *Land Degradation & Development*, 25(2), 173–183. doi: 10.1002/ldr.1158.

Stockmanna, U., Adamsa, M. A., Crawforda, J. W., Fielda, D. J., Henakaarchchia, N., Jenkinsa, M., and Zimmermannn, M. (2013). The knowns, known unknowns and unknowns of sequestration of soil organiccarbon (Review). *Agriculture, Ecosystems and Environment*, 164(2013), 80–99. doi: 10.1016/j.agee.2012.10.001.

Zauyah, S., Juliana, B., Noorhafizah, R., Fauziah, C. I., and Rosenani, A. B. (2004). Concentration and speciation of heavy metals in some cultivated and uncultivated ultisols and inceptisols in Peninsular Malaysia. *3rd Australian New Zealand Soils Conference*, pp. 5–9, December 2004, University of Sydney, Australia. Retrieved from: www.regional.org.au/au/asssi/(accessed June 6, 2016).

Zhao, H.-L., Cui, J.-Y., Zhou, R.-L., Zhang, T.-H., Zhao, X.-Y., and Drake, S. (2007). Soil properties, crop productivity and irrigation effects on five croplands of Inner Mongolia. *Soil and Tillage Research*, 93(2), 346–355. doi: 10.1016/j.still.2006.05.009.

Zingore, S., Murwira, H. K., Delve, R. J., and Giller, K. E. (2007). Soil type, management history and current resource allocation: Three dimensions regulating variability in crop productivity on African smallholder farms. *Field Crops Research*, 101(3), 296–305. doi: 10.1016/j.fcr.2006.12.006.

7 Soil Fertility and Management of Malaysian Soils

Zaharah A. Rahman, Amin Soltangheisi,
Aysha Masood Khan, Samavia Batool,
and Muhammad Aqeel Ashraf

CONTENTS

INTRODUCTION

Malaysian soils are highly weathered, acidic, and not very fertile for crop production. About 75% of these soils fall under the Ultisols and Oxisols soil group, which are normally found in tropical regions. Oxisols are more common than Ultisols in Malaysia where Ultisols constitute 9.3% and Oxisols 5.8% of the soils throughout the world (Fageria and Baligar, 2008).

Ultisols are usually acid soils, which have been formed under forest vegetation and are not natively fertile; therefore, they have to be enriched with lime and fertilizers for continuous agriculture. Malaysian Ultisols have the pH of <5. This low pH can profoundly reduce the microbial activity and consequently reduce the decomposition of organic matter, mineralization process, nutrient uptake, and eventually crop yield.

Oxisols are very highly weathered soils with the pH of below 5.0. Al toxicity and Ca, Mg, and P deficiencies are the major chemical constraints for normal plant growth in these soils. When the soil pH goes down to below 5, Al toxicity intensifies and becomes the major problem for crop production. These soils have the potential to become extremely productive with lime and fertilizer application.

Low cation exchange capacity (CEC), low pH, and low bases make Ultisols and Oxisols infertile for crop production. However, with proper management practices, good yield can be obtained, as in Malaysia, large areas of these soils are cropped to rubber and oil palm with desirable yields.

Granite, shale, schist, sandstone, basalt, andesite, and serpentinite are the major rock types in Malaysia. Weathering of these rocks under Malaysian conditions over a long period of time results in the formation of highly weathered minerals consisting of kaolinite, halloysite, gibbsite, geotite, and hematite in the clay fraction (Tessens and Shamshuddin, 1983). As a result of intense weathering, most of the plant nutrients are depleted from the soil profile, and therefore these soils have to be ameliorated with lime, organic matter, and other amendments.

MAIN LIMITATIONS TO CROP PRODUCTION IN SOIL MANAGEMENT

Productivity of the soil is improved by identifying the problems for the development of proper soil management techniques. Some features must be accounted for the proper soil and plant growth as explained in the following sections.

TOPOGRAPHY

It is considered as a more important factor in soil management as it determines the runoff and soil erosion potential of the area. Both can be a problem in the state. Due to steep slopes in Malaysia, soil erosion is of major concern. Particular attention must be given onto the drainage when erosion is noticed.

CLIMATE

It relates to low or high rainfall with or without dry period. Low rainfall generally supports a maximum crop yield by increasing photosynthesis in the sunshine hours. Heavy rainfall cause severe erosion and flooding in low-lying areas causing low yield due to loss of soil fertility.

SOIL EROSION

Serious problem in the tropics due to intense rainfall appears in the form of soil erosion and degradation. Potential for soil erosion can be high on hilly and steep slopes having loamy soils. Rain drops not only break soil aggregates but also carry away clay particles, leaving sand grains behind. Erosion takes away nutrients from the soil, so it can be termed as chemical erosion. Runoff water causes rills and gullies formation. This can be avoided by covering all road ends with vegetation, constructing silt traps and breaks, and by regular monitoring. Oil palm fronds minimize soil erosion and can also decreased by establishing terraces and cover crops.

FLOODING

In Malaysia during rainy season, flooding of low-lying areas may occur that can damage young oil palms or retard the growth of older palms. Matured oil palms are quite resistant to shallow flooding and moderately to high water table. However rubber does not tolerate flooding.

SOIL FERTILITY

High temperature and heavy rainfall in the tropical regions of Malaysia cause low fertility of soils. Most upland soils are developed over metamorphic, igneous or sedimentary rocks. Levels of nutrients, such as Ca, Mg, and minor elements such as Cu, B, Zn, can be monitored by leaf analysis coupled with field visits. High yield on such soils is achieved by adding fertilizers. Another common problem is Al saturation with related phosphate fixation, but fortunately oil palm and rubber can tolerate high Al saturation values.

HIGHLY WEATHERED SOILS OF MALAYSIA

In Malaysia, over a long period of time exposure of granite, shale, sandstone, basalt, andesite, schist, and serpentinite to weathering results in the formation of highly weathered materials in clay fractions dominated by gibbsite, goethite, hematite, kaolinite, and halloysite, generally termed as soil materials devoid of plant nutrients (Soil survey staff, 2010).

Feldspar dissolution in the water causes release of potassium through leaching into the groundwater, and further weathering of the clay mineral results in the formation of colloidal materials, which can be described as follows: (Duff, 1993)

$$6H_2O + CO_2 + 2KAlSi_3O_8 \rightarrow Al_2Si_2O_8(OH)_4 + 4SiO(OH)_2 + K_2CO_3$$

$$(feldspar) \quad (kaolinite)$$

$$H_2O + Al_2Si_2O_8(OH)_4 \rightarrow Al_2O_3.nH_2O + SiO(OH)$$

SOIL MANAGEMENT GROUPS

Different types of soils are found in whole Malaysia. To manage efficiency crops planted on these soils, soil mapping units are divided into different management classes. Most of the soils are mineral soils; brief description of these classes is provided in the following sections.

SOIL MANAGEMENT FOR CLASS 1

It consists of moderately deep (50–100 cm) to deep (>100 cm) well-drained soils, mostly well developed soils on flat, rolling, hilly, undulating, and steep terrains. Topsoil and subsoil comprise sandy clay loam facing soil erosion problems that can be controlled adding appropriate fertilizers, monitoring soil erosion, frond stacking, cover crops, and terracing.

SOIL MANAGEMENT FOR CLASS 2

It consists of well-drained, deep (>100 cm) to moderately deep (50–100 cm) soils present in undulating, rolling, hilly areas, steep terrains with sandy clay, loamy topsoil, and sandy clay in subsoil. Major problems concerned are soil erosion and nutrient removal, which are solved by proper soil erosion control especially on steeper terrain by terracing, frond stacking, cover crops, and by adding fertilizers.

SOIL MANAGEMENT FOR CLASS 3

It consists of poorly drained, less fertile, sandy to sandy clay loam-textured soil requiring high drainage, flood control, and P- and K-based fertilizers for proper functioning and development of crops.

SOIL MANAGEMENT FOR CLASS 4

It consists of all types of soil beds with shallow (50–100 cm), moderately deep (100–150 cm), deep (150–300 cm), very deep (>300 cm) organic soils lacking trace elements, poorly drained, flooding, and nutritional problems causing decline in the yield of oil palm after 12–15 years. Management of these soils needs suitable fertilizers, flood mitigation, and controlled drainage.

SOIL MANAGEMENT FOR CLASS 5

It consists of shallow (<50 cm), moderately deep (50–100 cm) to deep (>100 cm) soils with weakly or strongly cemented spodic horizon and sandy podzols. These soils are excessively drained with sandy texture, low moisture content, and low fertility. Seasonal flooding in soil with cemented spodic horizons occurs. Addition of fertilizer, empty fruit bunch (EFB), or organic mulching is necessary for the management of these soils.

SOIL MANAGEMENT FOR CLASS 6

It consists of soils with rock or laterite fragments at shallow depth (<50 cm) with weak root system resulting in poor anchorage, less nutrient retention capacity, wind damage, and soil erosion. These problems can be overcome by adding fertilizers with EFB or organic mulching.

Soil Management for Class 7

It consists of shallow (<50 cm), moderately deep (50–100 cm) fine sandy clay textured soil with limestone and coral limestone rich in Ca and Mg. Management of these soils can be done by adding EFB, and maintaining K level.

Soil Management for Class 8

It consists of shallow (<50 cm) lateritic soils with poor fertility, impediment to root penetration, wind damage, moisture stress, and fluctuations in yield. However, these soils are well drained. So proper fertilization, EFB mulching, and moisture conservation practices can well manage the soil.

Soil Management for Class 9

It consists of soil with acid sulfate layer below 50 cm depth, developed over marine and estuarine clays with poorly drained silty clay to clay textured soil. Management of these soils is also done by flood mitigation, by adding K-based fertilizers, and by drainage control.

Soil Management for Class 10

It consists of poorly drained soil developed over sulfide marine clay or brackish water deposits with silty clay to clay textured soil. Low nutrient uptake capacity and high root toxicity to acidity, Fe or Al are the main problems. Management is done by maintaining K level, water table around 60–80 cm depth, EFB, or bunch ash application.

Soil Management for Class 11

It consists of saline, unripe, and undrained potential acid sulfate soils. Management of these soils requires proper drainage.

Soil Management for Class 12

It consists of moderately deep (50–100 cm) to deep well-drained soil (>100 cm), highly erodible with low moisture retention capacity, fertility, and nutrient level resulting in fluctuations in the yield of crops. Management of these soils is done by adding fertilizers and by soil and moisture conservation.

ORGANIC SOIL MANAGEMENT

Organic soil in Malaysia is divided into nine main classes based on the problems concerned with the soil and management steps. Generally, these soils consist of shallow, moderately deep, deep, very deep organic soils with sapric and nonwoody material.

Most of these soils are poorly drained, have low fertility, and require high cost of construction. Yield of oil palm occurs after 5 years of planting, and yield declines after 12–15 years. Problems needed to be solved require good soil management practices with high planting density, good fertilizer programs with B, Cu, and Zn, water control, compaction of planting rows, thinning of stunted palms, flood mitigation, and constant monitoring for termite attacks.

ROLE OF ORGANIC MATTER IN MALAYSIAN SOIL MANAGEMENT

Organic matter is known as the nitrogen pool in soils. It supplies P and S for plant growth (Miller and Donahue, 1990). It also acts as a cementing agent to improve soil structure and to aggregate formation for protecting soils against erosion. Ultisols and Oxisols in Malaysia contain 1%–2% organic matter, which is not sufficient for plant growth. Most of these soils are cultivated under oil palm, rubber, and cocoa. As this area is large, it is unimaginable to apply organic matter to improve soil productivity, and the only way to keep organic matter at an optimal level is through good management practices such as zero burning. Organic matters are applied to field crops such as corn and groundnut to improve soil fertility and crop yield.

In highly weathered soils, there is a negative correlation between organic carbon and pH_0 (point of zero charge). pH_0 decreases one unit with every 1% increase in organic carbon (Gillman, 1985). Organic matter application in these soils can reduce pH_0 and consequently increase the pH of the soil and CEC. The increase in pH reduces Al toxicity for plant growth, and the increase in CEC prevents the leaching of basic cations in the areas with high rainfall and temperature.

The most common organic matters used by Malaysian growers are plant residues, compost, empty oil palm fruit bunches, palm oil mill effluents (POMEs), and chicken dung. Land degradation caused by soil acidification is one of the major problems in Ultisols and Oxisols of Malaysia. Soil degradation can be alleviated by the application of plant residues. Addition of plant materials can increase the soil pH and reduce exchangeable Al in soils (Hue and Amien, 1989). Mineralization of organic N results in increasing soil pH and if the C/N ratio is low, ammonium ions are released, leading to nitrification, which consequently makes the soil pH lower. Hence, we have to be cautious to treat the soil with these amendments. Rice husk compost (RHC) can be used as a source of N, P, and Mg. According to MARDI (2001), rice husk contains 0.4% N, 0.6% P_2O_5, and 0.2% MgO with the pH of 6.4. It can also reduce Al toxicity by increasing the pH and inactivating Al^{3+} with its phenolic and carboxyl groups. Chicken dung is the most commonly used organic fertilizer for vegetable farming in Malaysia. On average, it contains 1.6% N, 0.9% P, and 0.4% K (Seefeldt, 2013) and can supply these macronutrients for plants if applied at the appropriate rate. Malaysia is the second country after Indonesia in terms of palm oil production. Therefore, POME derived from digesting ponds of oil palm is too high and it is one of the common organic fertilizers used by farmers in Malaysia. It contains 2.20% N, 0.96% P, 0.93% K, 0.64% Ca, and some micronutrients such as Cu and Mn (Shamshuddin et al., 1998). The maximum rate of palm oil mill effluents (POME) application is 20 tonnes POME/ha, and its application in combination with ground magnesium limestone (GML) is recommended for Malaysia.

About a quarter of fresh fruit bunch (FFB) of oil palm is empty fruit bunch (EFB). Previously, this waste was incinerated, and the ash was added to the soil as an organic matter in oil palm plantation. It was recognized that this practice is not eco-friendly. Nowadays, this EFB is added directly to the soil to act as a mulch to keep the moisture and to improve the soil structure as well as to increase soil fertility. EFB contains substantial amounts of N, P, K, Ca, and Mg. Mulching with EFB also improved the soil exchangeable K, Ca, and Mg and pH (Lim and Zaharah, 2002). The frequency of fertilizer application did not affect the leaf nutrient levels and the yield of oil palm. Thus, applying EFB and supplementing it with N and K fertilizers should always be advocated to exploit their beneficial combined effects on oil palm nutrition and yield.

Another organic matter, which is used in oil palm plantation is old fronds. These fronds are cut from time to time during FFB harvesting to keep the trees clean healthy. Then, they were put between the oil palm rows or along the harvesting path for decomposition. After decomposition and mineralization, they release nutrients, which are taken up by the growing oil palms. These two practices can cut down fertilizer cost in oil palm plantation and ameliorate the soil structure and oil palm nutrition.

Rubber trees are felled after about 25 years of continuous production. Immature phase usually takes 3–4 years. During this phase, growers commonly plant field crops such as corn in the interrows to earn some income while waiting for rubber trees to mature. At this time, they also use some amendments such as POME. This can help the current field crop to produce more yields and also add extra nutrients to the soil for better and faster growth of rubber. As a result of such a practice, rubber trees can be tapped earlier than the normal age of about 4.5 years.

One of the most innovative technologies in organic matter management is zero-burning technology. It is now being practiced by oil palm and rubber estates in Malaysia during replanting phase. Rubber and oil palm are replanted when it is due, usually after 25 years of continuous production. In this technique, the plants are felled, shredded, stacked, or buried underground to decompose naturally. An oil palm biomass from 1 ha of land contains 642 kg of N, 58 kg of P, 1384 kg of K, and 156 kg of Mg (Che Johari, 2005). The amounts of nutrients contained in 1 ha of natural rubber land are 72 tonnes of C, 999 kg of Ca, 826 kg of N, 767 kg of K, 154 kg of Mg, and 107 kg of P (Yew, 2000). About 44% of these nutrients would be returned to the soil. This practice, therefore, reduces the fertilizer cost tremendously.

A range of green manures can be grown on farm to supply plant nutrients and organic matter for the enrichment of the soil for the next cultivation (e.g., herbaceous legumes grown in a crop rotation, or agroforestry systems where the leaves and stems of nitrogen fixing trees/shrubs are used to fertilize adjacent crops). Legume tree prunings have been shown to improve soil physicochemical and biological properties by directly increasing soil organic matter and by providing available plant nutrients through mineralization (Bah and Zaharah, 2001; Chintu and Zaharah, 2003).

The use of organic matter for ameliorating acid sulfate soils did not get much attention in Malaysia as compared to normal acid soils or highly weathered soils. The reason is that the soil has high content of organic C. Many reports have shown that humic acid, as well as other low molecular organic acids, plays an important role in the detoxification of Al (Tan and Binger, 1986; Hue et al., 1986; Hue, 1992). The effectiveness of organic acids to chelate Al depends on their chemical reactivity. Based on the chemical reactivity, organic acids can be differentiated as organic acids with acidic characteristics attributed only to the presence of carboxyl group, and organic acids with acidic characteristics attributed to the presence of both carboxyl and phenolic groups. The first group may exhibit some complexing capacity through the acidic effect or electrostatic attraction. The latter, however, have complexing capacity through attraction, coadsorption, complexing, and chelating reaction (Stevenson, 1994). It is more likely that organic acids of most of acid sulfate soils fall into the first group. Adding organic manure in acid soils may substitute for lime as a soil amendment to increase soil pH and decrease the total soluble Al in soil solution.

ROLE OF LIMING MATERIALS IN MALAYSIAN SOIL MANAGEMENT

Seventy two percent of Malaysian soils are occupied by highly weathered acid soils (Ultisols and Oxisols; IBSRAM, 1985). These soils have high acidity, low effective (ECEC), and high aluminum saturation throughout the soil profile (Tessens and Shamshuddin, 1983). Aluminum and manganese toxicities, calcium, magnesium, phosphorus, and micronutrient deficiencies, and water stress arising from subsoil acidity are the most important concerns for crop production in acid soils. In most of the cases, the soils are so acidic that liming is necessary to obtain economical yields in annual crop production.

The pH of Oxisols and Ultisols in Malaysia is low, ranging from 4 to 5, and it is generally lower in Ultisols as compared to Oxisols. The most important exchangeable cations for plant nutrition and soil fertility are Ca, Mg, and K. Ca and Mg are macronutrients, which are needed in large amounts for normal plant growth. The Ultisols and Oxisols in Malaysia are inherently deficient in these cations. Ground magnesium limestone (GML) is used for liming to overcome this nutrient deficiency. Exchangeable Al is moderately high in the Ultisols and Oxisols in Malaysia and may cause Al toxicity especially in crops, which are intolerant or Al sensitive such as maize and cocoa. Lime application can modify Al concentration to normal levels for the optimum plant growth.

Kaolinite and sesquioxides (oxides of Fe and Al) are the two dominant minerals in clay fraction of the Ultisols and Oxisols of Malaysia (Tessens and Shamshuddin, 1983; Paramananthan, 2000). These minerals have negative charges or even some of them such as oxides of Fe have net positive charges. The consequence of such a combination is the low CEC of Ultisols and Oxisols. The values of CEC in the Ultisols and Oxisols of Malaysia are usually less than 10 $cmol_c$/kg soil. Sesquioxides content in Oxisols is higher than Ultisols, which resulted in the lower values of CEC in Oxisols as compared to Ultisols.

Ca, Mg, and K are the most important macronutrients, which are needed by plants in large amounts for continual growth. Malaysian growers commonly apply fertilizers to supply these nutrients for plant growth. Soils with low CEC do not have the ability to absorb these basic cations, and therefore, they are mostly lost into the groundwater via leaching. This problem can be overcome by lime or organic fertilizer application.

Liming reduces the possibility of Mn^{2+} and Al^{3+} toxicity. It also improves microbial activity, physical condition, symbiotic nitrogen fixation by legumes, palatability of forages, and the availability of some nutrients such as phosphorus and molybdenum. Lime is an inexpensive source for Ca^{2+} and Mg^{2+} when these nutrients are deficient at lower pH.

Oil palm is acid tolerant and can grow at the soil pH of 4.3 and at high Al concentration (Auxtero and Shamshuddin, 1991). Hence, Malaysian Oxisols and Ultisols are suitable for oil palm cultivation, and liming is not necessary as the area under oil palm is very large and liming is not economical. Like oil palm, rubber is also acid tolerant and can grow at the soil pH of 4–5. Therefore, there is no need to apply lime onto the soils for rubber cultivation.

The best pH for high-quality cocoa plantation is 6.0–6.5. In Malaysia, the soils with this range of pH are rarely found; hence for cocoa plantation, liming is required. According to MARDI (1990), lime has to be applied at the rate of 20–30 g GML per bag during the nursery stage. Moreover, it is suggested that 150 g of GML should be applied for each cocoa tree during the time of planting (Othman, 1993). Furthermore, for the first, second, third, fourth, and fifth year, the rate is 200, 200, 340, 340, and 340 g/tree, respectively.

Mn^{2+} activity is higher in Oxisols as compared to Ultisols and may cause Mn toxicity in cocoa plants that are grown in these soils. Foliar application of Ca fertilizers can result in balance of nutrients in the plant system and can reduce the Mn toxicity within the plant. Although this method is more expensive, it is more effective than GML application.

LIMING MATERIALS

Agricultural liming materials are calcium and/or magnesium containing materials, which are capable of increasing pH in acid soils. These materials include dolomitic lime, calcitic lime, slaked lime, basic slag, and wood ash as given in Table 7.1. The most common limestone in Malaysia is dolomitic limestone, which is locally known as GML.

TABLE 7.1
Liming Materials, Their Compositions, and Neutralizing Power

Liming Material	Formula	Neutralizing Power	Characteristics
Dolomitic lime	$CaMg(CO_3)_2$	95–109	Contains 78–100 g/kg of Mg, 180–210 g/kg of Ca
Calcitic lime	$CaCO_3$	100	Contains 284–320 g/kg of Ca
Slaked lime	$Ca(OH)_2$	136	Fast reacting
Basic slag	$CaSiO_3$	86	By-product of iron industry
Wood ash	Variable	30–70	Caustic and water soluble

Source: Fageria, N.K. and Baligar, V.C., *Adv. Agron.*, 99, 345–399, 2008.

USE OF BASALT AS SOIL AMELIORANT

Basalt, a naturally occurring rock, contains essential elements such as Ca, Mg, K, P, S, and Fe, which are beneficial for plant growth. It can be used as a soil amendment for nutrient poor acid soils of the tropics. Anda et al. (2015) demonstrated that basalt application can increase the soil pH, base saturation, and CEC and alleviate Al and Mn toxicity of an Oxisol (a fine, clayey, kaolinitic isohyperthermic, Rhodic Hapludox, Segamat series).

Anda et al. (2013) showed that combined application of crushed basalt with rice husk compost (RHC) has positive effects on increasing the soil organic C functional groups (O-alkyl, di-O-alkyl, and carboxyl C), along with the generation of aromatic, alkyl, and methoxyl C as new functional groups, which play an imperative role in producing soil negative charges in Oxisols, which helped to increase the Ca, Mg, K, Na, and Si in the soil resulted in the improvement of cocoa growth.

MANAGEMENT OF MAJOR CROPS GROWN IN MALAYSIA

OIL PALM

Oil palm is the most important agricultural crop in Malaysia. About 5.4 million ha is currently under oil palm cultivation in this tropical country. Sabah has the largest plantation area of any single Malaysian State. The soil in Sabah is fertile and can produce the highest yield as compared to other states. Oil palm is grown on upland areas, which had earlier been cultivated under rubber or coconut, but very large areas of primary forest are also used. Most of these areas are where Oxisols and Ultisols occur. It can grow very well with good soil management practices. NPK fertilizers have to be applied to the soil for oil palm cultivation. At least 50% of the total fertilizer consumption in Malaysia is used in the oil palm industry (Chew et al., 1992). The cost of fertilizing is around one-third of the total production costs of a plantation (Caliman et al., 2001).

Nitrogen fertilizers have to be applied according to fertilizer experiments and leaf analysis. Ammonium sulfate is the standard nitrogen fertilizer, although it can acidify the soil as the ammonium ion is oxidized to nitrate. This is not a serious problem unless it is used continuously for a long time at heavy rates. This fertilizer has an antagonistic interaction with magnesium resulting in magnesium deficiency in oil palms.

Urea hydrolyzes to ammonium bicarbonate and then to ammonia, which can be volatilized, and therefore, it shows a low efficiency as compared to other ammonium salts. It produces from local gas supplies and the manufacturing cost is cheap; hence, it is used extensively.

The main sources of phosphorus in oil palm plantation are phosphate rocks (PRs), triple superphosphate (TSP), and diammonium phosphate (DAP). The water soluble TSP source is commonly applied to palms at the immature stage in the form of compound fertilizers. However, by far, the largest usage of P fertilizer is in the form of PRs, which are effective on the acid soils and more economical for use in the mature palms. It is a slow release fertilizer and dissolves slowly under low pH conditions. The PRs vary widely in physical, chemical, mineralogical, and reactivity characteristics. For application on oil palm, PRs are finely ground for two reasons. First, the soils are acidic, and PR dissolves rapidly at the soil pH of below 5.5. Second, oil palms can absorb P slowly during the year (Corley and Tinker, 2008). Phosphate can react with oxides and hydroxides of Fe in the soil and consequently increase the negative charges and the pH of the soil (Tessens and Zaharah, 1983). Although this phenomenon improves the productivity of the soil, it fixes P and makes it unavailable for uptake by oil palm.

P fertilizer requirements range from 0.3–0.4 kg P/palm (coastal clay soils) to 0.4–0.7 kg P/palm (inland soils) during the first 3 years of growth. Phosphate fertilizer is broadcasted in the weeded circle of young palms (Goh and Härdter, 2003). The application may be decreased after this time as a result of the smaller requirement of P for mature palms in which the nutrient is already provided from the soil. Potassium chloride is the standard K fertilizer in Malaysia; however, it is the most expensive fertilizer applied.

Oil palm yield on Ultisols and Oxisols of Malaysia is 20–30 tonnes FFB/ha/year, which can be increased up to 40 tonnes FFB/ha/year with planting high-yielding oil palm clone.

The beneficial age of oil palm is 25 years, and after this time they have to be replanted. At this age, they are not productive anymore, and also they are too tall for harvesting FFBs using standard practice. Zero-burning technology is used in this stage. The fronds and trunks are cut and chopped into small pieces and are placed on the ground as shown in Figure 7.1. After mineralization, essential nutrients are released into the soil and hence will reduce the fertilizer application and cost.

RUBBER

The first rubber estate in Malaysia was established in Melaka in 1903. Malaysia is the biggest consumer of pure latex and fifth in the consumption of natural rubber in the world. Asia is the biggest producer of natural rubber, led by Thailand, Indonesia, and Vietnam. Malaysia is the third country in terms of latex production after Thailand

FIGURE 7.1 Oil palm trunk and fronds chipped and spread in rows on the ground before planting a new crop. (From Ooi, L.H., Agronomic principles and practices of replanting oil palm, in *Seminar on Agronomic Principles and Practices of Oil Palm Cultivation*, eds. Goh, K.J., Chiu, S.B., and S. Paramananthan, Agricultural Crop Trust, Sibu, Malaysia, pp. 171–197, 2011.)

and Indonesia. Not only the latex but also timber from the trees are the yield of rubber and have a very high demand in the marketplace. The current area covered by rubber is estimated to be about 1.2 million ha distributed throughout the country. Most of rubber is grown on Ultisols and Oxisols. This plant is acid tolerant and can grow efficiently on soils with well drainage systems.

Malaysian soils, which have been under rubber cultivation for some time, are generally deficient in nitrogen. Ammonium sulfate $((NH_4)_2SO_4)$ is the principal nitrogen fertilizer applied for rubber cultivation in Malaysia. It is highly water soluble and the ammonium from this fertilizer undergoes nitrification, which release H^+, and hence, its long-term application would increase soil acidity slightly. However, rubber is acid tolerant, and the reduction in pH would not affect its growth much. This fertilizer supplies not only N but also S, which is a macronutrient for normal growth of rubber. Uptake of ammonium sulfate by young plants is rapidly accomplished.

Most Malaysian inland soils contain relatively small quantities of available phosphate, and young rubber trees growing in these soils give a marked response to phosphate applications in their early years. The preferred phosphorus fertilizer in rubber estates is phosphate rock, which has been used since many years ago, and Christmas Island rock phosphate is the most common one. It contains 38% P_2O_5 of which 11% is acid soluble.

When phosphates are applied to the soil, much of them combine with iron and aluminum to form relatively insoluble compounds, which cannot be utilized by plants. The degree of fixation, which occurs, varies with soil type and acidity and is roughly the same for soluble as for insoluble phosphate. The more acid the soil the quicker the fixation: numerous experiments have shown that phosphates applied broadcast to soils in Malaysia do not penetrate more than a few centimeters, and in some of the high fixing soils, about 90% of the amount applied may be held in the top 3 cm of soil. Application of a nitrogenous fertilizer reduces the degree of phosphate retention, and ammonium sulfate increases the availability of native phosphate.

Muriate of potash is the form in which K is usually supplied as fertilizer to rubber in Malaysia and contains 60% K_2O. Sulfate and nitrate of potash are other forms containing 26%–27% K_2O and 10% MgO.

Similar to oil palm estates, rubber estates are practicing zero-burning technology without fail, which puts Malaysia at forefront in promoting eco-friendly technology for protecting the environment (Othman Yacob and W.H. Wan Sulaiman). This technology is Malaysia's contribution to the world in the reduction of global warming. Important part of soil management for rubber includes legume and cover crops, providing effective cover against soil erosion in newly and replanted rubber. Maintaining a grood cover crop with regular fertilizer application up to third or fourth year from planting will reduces the demand for nitrogen fertilizer, protect soil, and also increase soil fertility. All these management practices represent basic principle for cropping system management in humid tropics. In Malaysia and other humid regions to maintain soil fertility and retain crop productivity, soil must be moist, cool, covered, and undisturbed.

COCOA

Cocoa is the third major commodity crop after oil palm and rubber in Malaysia. In the 1970s, Malaysia was the sixth biggest cocoa bean producer in the world with more than 400,000 ha under cocoa cultivation. Since then, many cocoa plantations are abandoned or replaced with oil palm in a way that the area under cocoa plantations is about 50,000 ha nowadays, and the production is even not enough to feed the local factories producing cocoa products in the country.

Cocoa needs an annual rainfall of 1250–2800 mm with an average temperature of 18°C–32°C and without any drought exceeding 3 months. Areas with the rainfall of more than 3000 mm such as Johor are not suitable for cocoa plantation because under this condition, cocoa can be infested by diseases, which lead to the yield reduction. Cocoa can grow well on soils with good drainage and fertility with soil pH of 4.5–6.5 (Hardy, 1971). It can, therefore, cope with both acid and alkaline soil, but excessive acidity (pH 4.0 and below) or alkalinity (pH 8.0 and above) must be avoided.

The composition of 50% silt, 30%–40% clay, and 10%–20% silt would be suitable with C:N ratio of 10–12, organic C of more than 3%, CEC of 3–15 cmol$_c$/kg soil, available P of more than 15–20 mg P/kg soil, exchangeable K of 0.25 cmol$_c$/kg soil, and low exchangeable Al would be suitable for cocoa cultivation. Sabah is the best growing area in Malaysia for cocoa production in terms of soils and climatic conditions.

Soil pH can be increased by liming or ground basalt application. Basalt can also supply Ca, Mg, K, P, and S for cocoa plants; however, it needs a long time to disintegrate and dissolve completely. The application of basalt with organic fertilizer can supply N needed by cocoa for its healthy growth. Negative charges on the exchange complex of the variable charge minerals increase with increasing pH of the soil and consequently retain basic cations in the topsoil for uptake by cocoa plants.

Table 7.2 shows an estimation of nutrient recommendation based on 1075 trees/ha of cocoa plants. Fertilization program for cocoa depends on various factors such as soil fertility, rainfall pattern, growth pattern, yield potential, main fruiting season, and pruning schedule as presented in Table 7.3.

TABLE 7.2
Estimation of Nutrient Requirements of Cocoa Plants at Different Stages of Development from Whole Plant Analysis

Stage of Plant Development	Range of Age of Plants (months)	Average Nutrient Requirement in kg/ha						
		N	P	K	Ca	Mg	Mn	Zn
Seedling (nursery)	5–12	2.4	0.6	2.4	2.3	1.1	0.04	0.01
Immature (field)	28	136	14	151	113	47	3.9	0.5
First year production (field)	39	212	23	321	140	71	7.1	0.9
Mature (field)	50–87	438	48	633	373	129	6.1	1.5

Source: Thong, K.C. and Ng, W.L., Growth and nutrients composition of monocrop cocoa plants on inland Malaysian soils, in *Proceedings International conference on Cocoa and Coconuts, 1978,* Incorporated Society of Planters, Kuala Lumpur, Malaysia, pp. 262–286, 1980.

TABLE 7.3

Fertilizer Program Based on Different Rainfall Pattern, Flowering, and Fruiting Seasons in Malaysia

Fertilizer Application Period	Fertilizer Rate (g/plant)			Purpose
	1[a]	2[a]	>3[a]	
A. (Johor, Malacca, Perak) After short dry season in early of the year (March–April)	170 CF or 100 urea or 175 AS and 200 PR	250 CF or 150 urea or 250 AS and 250 PR	350 CF or 300 urea or 500 AS and 250 PR	To promote flowering
Liming[b] (1 month after first fertilizer application)	300–500	300–500	300–500	To reduce soil acidity and increase soil absorption capacity
3–4 months after first fertilizer application with or without second short dry season	170 CF or 100 urea or 175 AS and 75 MP	250 CF or 150 urea or 250 AS and 175 MP	350 CF or 300 urea or 500 AS and 250 MP	To promote pollination and flowering
3–4 months after second fertilizer application (October–November)	170 CF or 75 MP	250 CF or 175 MP	350 CF or 250 MP	To promote fruiting, which will be harvested in the early of flowering year
B. (Selangor, North and East Coast) After long dry season in early of the year (April–May)	170 CF or 100 urea or 175 AS and 200 PR	250 CF or 150 urea or 250 AS and 250 MP	350 CF or 300 urea or 500 AS and 250 PR	To promote flowering
Liming[b] (1 month after first fertilizer application)	300–500	300–500	300–500	To reduce soil acidity and increase soil absorption capacity
3–4 months after first fertilizer application (July–August)	170 CF or 100 urea or 175 AS and 75 MP	250 CF or 150 urea or 250 AS and 175 MP	350 CF or 300 urea or 500 AS and 250 MP	To promote fruiting

Source: MARDI, *Panduan Menanam Dan Pemprosesan Koko*, Berita Publishing Press, Kuala Lumpur, Malaysia, 1990.

CF, compound fertilizer; AS, ammonium sulfate; PR, phosphate rock; MP, Mutriate of Potash.

[a] Fruiting year.

[b] Limestone is used for coastal area, while GML are used for alluvial area.

TECHNIQUES AND TOOLS TO BUILD WELL-MANAGED SUSTAINABLE SOIL

Some tools to build and make amendments in the management of sustainable soil include animal manure, composting farm manure, organic materials, cover crops and green manure, humates, tillage reduction, minimization of nitrogen fertilizers, rotation, and continuous monitoring to indicate success or failure.

SOIL WATER MANAGEMENT

Irrigation system without water cannot exist. Plantation needs excess water supply along with other soil management practices. Water management practices include irrigation of land and drainage and soil conservation practices, which gives good response. Soil water table is maintained at 75 cm from ground surface. Areas with poor rainfall, excessive drainage of soil, soil causing restricted rooting and rainfall less than 1700 mm/year are suitable for soil moisture conservation measures and irrigation.

SOIL ACIDITY MANAGEMENT

Acidity of soil mostly occurs in Malaysia. Two factors are mainly involved: one, the soil itself is acidic, and the other management practices such as choice of nitrogen fertilizers make it acidic. Cocoa is quite sensitive to low pH, but rubber and oil palm, they are quite resistant. If soil becomes too acidic to be detrimental to plants, mulching, and liming will increase the soil pH.

SOIL FERTILITY MANAGEMENT

In Malaysia, fertility among various soils varies from area to area, indicating that they are not only site specific but also crop specific. Fertilizer requirement depends on inherent soil nutrient status. For nutrient requirement diagnosis, foliar analysis is a supplementary tool. The more efficient the fertilizer, the lower will be the risk of manuring on the environment. Sandy soils of Holyrood and Malau series required higher frequency with smaller dressings of soluble fertilizers to immature series soils. Actual frequency of the fertilizer depends on tree age, ground conditions, type of fertilizers, and rainfall and crop requirement. High rainfall as in humid tropic causes nutrient loss.

SOIL AND WATER CONSERVATION MANAGEMENT

It is one of the necessary factors in soil management. Soil erosion and runoff results in water loss with breakdown of soil particles in topsoil due to rain. Excess rainfall boosts rate of infiltration into the soil. Main objective of soil and water conservation is to obtain maximum production in a specific area without soil degradation and environmental pollution.

MANAGEMENT OF DIFFERENT TYPES OF SOILS

MANAGEMENT OF SHALLOW LATERITIC SOIL

In Peninsular Malaysia, Malacca and Gajah Mati series belongs to shallow lateritic soils. Oil palm root and plant growth are restricted on such soils due to low effective soil volume, low water holding capacity, low CEC, high P-fixing capacity and poor nutrients. Productivity on lateritic soils can be increased by water and soil management and by improving soil fertility. Manuring is done to increase the fertility of soil, and phosphate rock provides P for strong root activity. These activities reduce runoff and soil erosion and build up organic matter in the soil.

MANAGEMENT OF SHALLOW SALINE SOIL

Generally, low rainfall regions of Malaysia have saline soils. Plants normally cannot withstand with high salt content in the soil. Some Malaysian companies such as K.L. Kepong, golden hope plantation Bhd. and Sime Darby Bhd. have successfully grown oil palm on this soil. These soils mostly occur by the sea or around the river mouth. Preventing intrusion of the sea or brackish water into the land by constructing bunds 3 m above the highest tide level. After this drainage networks are laid down to reduce water table. Periodic flushing of the soil for oil palm plantation continues for 4 years to reduce soil conductivity to below 2000 μmhos/cm within 45 cm of the soil, and then field drains remove water from 50 to 70 m from soil surface. These management services are necessary to grow oil palm.

MANAGEMENT OF SHALLOW ACID SULFATE SOIL

In Peninsular Malaysia, most of the acid sulfate soils are found in the coastal plains of the west coast. Some acid sulfate soils are also found in the coastal plains of Sarawak and Sabah. It is estimated that 110,000 ha in Peninsular Malaysia is covered by acid sulfate soils, whereas for the whole country, the size is about 0.5 million ha (Kanapathy, 1973). These soils have low pH of less than 3.5, and contain yellowish jarosite mottles. About 20,000 ha area are grown with oil palm. Such soils possess problems such as water logged and high acidity. Careful management of these soils is required. Liming shows no effect in acidity reduction. Prime requirement is management of water table to be above the pepsitic layer as long as possible and periodic flushing of the soil to remove toxic polyvalent ions (Al^{3+}) and highly acidic water.

MANAGEMENT OF DEEP PEAT SOIL

In Sarawak, Malaysia, 1.5 million ha is covered by peat soil. Peat is a material having a loss of ignition of above 65%. Organic soils with the thickness of 50–150 cm are classified as shallow peat, whereas those with the thickness of more than 150 cm are deep peat (Tie, 1982). Major crop grown on this soil is oil palm. Peat soils generally have high-water holding capacity, poor aeration, and low-bulk density. If drainage occurs, soil becomes hard and dry. Management of this soil needs perimeter drains to remove excess water, but not over drain. If occurs, it adversely affects establishment and plant growth. Periodic flushing of acidic and storm water is also carried out. High nitrogen concentration must be applied initially and must be reduced during mature phase. Mineralization releases P into the soil with low Al and Fe for fixation. These soils are deficient in potassium, copper, zinc, and boron and must be added in initial stage. Growth and yield of the plants can be better improved by better understanding the physical and chemical properties of the peat such as mineralization rate, better cationic–anionic balance, and increasing pH.

MANAGEMENT OF PODZOL OR SPODOSOLS

Most of these soils occur in beach ridges interspersed with swales (BRIS) soils in the east coast of Peninsular Malaysia and on moderate hills in East Malaysia. Major crops includes vegetables, cashew nut trees, star fruit trees, and tobacco. Soil management requires moderate water drainage, suitable fertilizer input, high reactivity phosphate rocks, and large amount of GML to prevent Mg deficiency.

MANAGEMENT OF SANDY SOIL

Such soils occur besides mining or ex-mining areas and flat river basins. Major classes of these soils are Subang, Nangka, Sg. Buloh, Lintang, and Jambu series. Management practices on these soils show requirement similar to podzols with some differences.

MANAGEMENT OF FLOODED RICE SOILS

Rice is the third most cultivated crop in Malaysia after oil palm and rubber. The area under paddy rice fields is estimated around 692,340 ha in 2012 dominated by Peninsular Malaysia (76%) and remained constant during the last 30 years. Malaysia's mean rice productivity is only 3.3 tonnes/ha/year, which is very low as compared to Australia, the world's most efficient producer of rice, which produces an average of 8.7 tonnes/ha/year. Singh et al. (1996) estimated that the potential yield of paddy in Malaysia is 7.2 tonnes/ha/year. According to the Department of Agriculture in Malaysia, the country produced 1,847,000 tonnes rice in paddy fields in 2013. In order to increase the self-sufficiency level (SSL) of rice in Malaysia, which is now 73%, improving yield efficiency of rice seems to be the only alternative as the good and fertile lands are scarce, and expanding the rice cultivation area seems impossible. Also there is a great potential to use acid sulfate soils in Malaysia for paddy cultivation and the management of these soils was discussed earlier.

There are 0.3 million paddy farmers in Malaysia of whom 40% are full-time farmers. The farm size of 65% of the paddy farmers is below 1 ha (Md. Mahmudul et al., 2010). Paddy farmers can be categorized as four groups: (1) tenant farmers who rent the land for farming, (2) farmers as well as the land owners who farm on their own land, (3) combination of (1) and (2) where these farmers not only farm on their own land but also rent other people's land, and (4) land owners who lease their land to others for farming.

In Malaysia, the rice fields are usually ploughed twice before sowing or planting. After water is introduced, a round of puddling and land leveling will be done. Crop establishment can be done either by direct seeding, which is broadcasting pregerminated rice seeds directly into the field manually or using row seeder, or transplanting, which is planting 25–35-day old seedlings into the field manually or by machines using seedlings sown on trays.

Many rice fields in the coastal plains of Malaysia are still facing the threat of soil salinization, especially during high tide. Kimi and Daham (1993) suggested that simple flooding and flushing of the soil with the simultaneous application of lime can reduce the salt effect and can increase the rice yield.

About 3.5%–13.5% of rice grains are lost during harvest as reported by MARDI (2007) because of low efficiency of threshing mechanism.

There are some areas with soft soil under paddy cultivation. These soils are formed as a result of using unsuitable farm machinery, low drainage density, climate change, and planting out of the schedule. Areas with soft soil can cause serious issues in paddy fields especially for mechanization. Machines are used in all different steps of paddy cultivation including land preparation, transplanting, land leveling, crop maintenance, and harvesting. Farmers have to allow these soils to dry or lay idle for sometime and the standing water to drain out and the subsoil to dry and harden.

SOIL MANAGEMENT IN PENINSULAR MALAYSIA

MANAGEMENT OF ULTISOLS AND OXISOLS IN PENINSULAR MALAYSIA

Ultisols and Oxisols, in Peninsular Malaysia are mainly utilized for rubber and oil palm cultivation. The soils are acidic with high Al saturation, low CEC, exchangeable bases, and available P. Acidity of these soils can be decreased by adding GML. Experiments conducted using maize and peanuts grown as intercrops in immature rubber plantings showed GML effect on an Ultisol (Bungor soil series) shows higher yield with optimum rate of 2 tonnes/ha and 1 tonne/ha on an Oxisol (Munchong soil series). Higher amount of GML causes the reduction of Zn, Cu, and Mn in the soil. More than 90% of the peanut on both soils is obtained by applying GML of 0.5 tonnes/ha. Applying 1 tonne/ha GML on the Bungor soil increased the pH up to 5 with no effect on Munchong soil (Sharifuddin et al., 1989).

Increasing demand for latex and timber in Malaysia include Malaysian Rubber Board (MRB) to conduct research on rubber to enhance its productivity. Ultisols are considered as first class for rubber production with high yield. Soil management for rubber in Malaysia is mainly done on Ultisols by adding nitrogenous fertilizers. Ammonium sulfate (21% N), a white crystalline chemical, is largely applied on inland soils but due to its high cost, an alternative Urea, classified chemically as an organic compound containing 46% N by weight, is applied as it works equally well providing same amount of nitrogen at equivalent of 10% N to the soil in its efficiency and growth, increasing growth, yield, and green color of leaves. It is cheaper than ammonium sulfate, extensively added with controlled dosage especially at nursery level. A local company, PETRONAS, has launched its own brand of urea fertilizer, Agrenas, which is expected to give higher yield and to support the growth of Malaysia's agriculture sector. It can be concluded that the best fertilizer treatment for raising RRIM 2000 series rubber scions is applying urea or ammonium sulfate at a rate equivalent to 10% N (Mokhatar and Daud, 2011).

Oil palm grown on Kuala Brang series soil (an Ultisol) was tested for its proper management and fate of phosphorous, where phosphorous was added in the form of

Christmas Island phosphate rock (15% P) for 17 years consecutively. Average soil pH for zero phosphate treatment was 4.45 but was increased after the addition of phosphate rock due to the presence of Ca, causing enhancement in the base saturation of the soil resulting in high pH at a depth of 10 cm, high CEC values with increasing rates of P, and low Al-P fractions. So a better management of soil is done (Zaharah et al., 1985).

MANAGEMENT OF ORGANIC SOIL

Organic soil that is widely spread in Malaysia consists of woody materials, high CEC, high C/N ratio, low nutrient contents, and acidic reaction. Common problems associated with this soil in Malaysia is micronutrient deficiency especially of copper, zinc, manganese, and iron. Deficiency of copper and zinc is compensated by foliar spray of copper sulfate and zinc sulfate solution. The soil of Department of Agriculture (DOA) peat station, Sg. Burong, Tg. Karang, Selangor is modified to utilize peat soil for the cultivation of crops, mango, coconut, coffee, tapioca, and vegetables. The color of the soil is dark brown (7.5 YR 3/2) with more than 30% organic matter to a depth of 60 cm and with water table at about 15 cm below the surface, but the area floods during the rainy season, and organic matter in the soil is completely decomposed (sapric material). The soil has undergone repeated liming and drainage resulting in the increasing soil pH up to 5 and in accumulation of bases for proper crop production. At 45–60 cm depth, buffering capacity is very high with a pH of 4–5. A liming rate of about 3 tonnes/ha is recommended to increase the pH to 5.5 in the top 30 cm of the soil. Excess liming and fertilizing should be avoided as it can cause a nutrient imbalance. Crops normally should grow on specially constructed bunds. In this way, excess salts are leached to the lower horizons (Shamshuddin et al., 1985).

MANAGEMENT OF ACID SULPHATE SOILS

The soils in the west coast of Malaysia significantly contain clayey deposits in the form of pyrites (FeS_2) of marine and brackish water origin that often grades into the peat swamps. About 0.5 million ha of acid sulfate soil occurs in Malaysia. Soils of Kemasin-Semerak Integrated Agriculture Development Project (IADP), Kelantan belong to the Parit Botak series classified as a Typic Sulfaquepts, managed for commonly grown crops such as oil palm, cocoa, coconut, rubber, and rice. These soils are mostly enriched with sulfur in the form of sulfide, which is generally recognized by rotten egg smell. Acid sulfate soils form an important group of soils in Peninsular Malaysia with pH less than 3.5, which decreases with depth and high Al. When this pyrite is exposed to the atmosphere, jarosite [$FeK_3(SO_4)_2(OH)_6$], a new straw-yellow colored (2.5 Y 8/6) mineral is formed. As a result of pyrite oxidation, high amount of acidity is released into the soils and environment. It is difficult to grow crops on these soils. However, there are plant species that thrive well under acid sulfate soil conditions, including gelam (*Meluleuca leucadendron*), nipah (*Nipa frutescens*), and mangrove (*Rhizophora mucronata*) species. To ameliorate the infertility of acid sulfate soils for rice cultivation, GML, hydrated lime, liquid lime, and fused magnesium phosphate and basalts under submerged and moist condition can be applied.

CHANGES AND EFFECTS ON SOIL UNDER DIFFERENT CONDITIONS

CHANGES IN SOIL UNDER SUBMERGED CONDITION

Lime application under submerged condition stimulates the releases P to the soils. The soil and water pH increased from 3.43 to 5.01 and from 2.38 to 5.17, respectively. It required 6 tonnes GML/ha to increase the pH to about 5. Submerging the soil also decreases the acidity. At this rate of GML application, exchangeable Al was decreased to <1 $cmol_c$/kg soil after 12 weeks due to oxido-reduction processes in the soil during the incubation period. These soils also contain high concentration of Al^{3+} and Fe^{2+} causing low pH problems and nutrient deficiency. High Al concentration causes the retardation of root elongation in rice. These specific areas are designated as sulfuric horizons. Four soil series, namely, Kuala Perlis, Parit Botak, Sedu, and Guar are identified to have sulfuric horizons within 50 cm of the surface and in Jawa, Telok, Jawa (shallow) and Tong Kang series is identified to be within 50–100 cm. Hydrated lime is also applied to increase the pH and to decrease the exchangeable Al at a rate of 4 tonnes/ha but with no remarkable decrease in Fe. Another way to treat the soil under submerged condition is the addition of liquid lime. This application causes no significant rise in the pH even at higher concentration but positively decreases exchangeable Al and the exchangeable Fe is erratic. Currently, fused magnesium phosphate (FMP) at a rate of 125 kg/ha is applied. This dose not effectively increases the pH up to 5 but also decreases exchangeable Al within 6 weeks (Suswanto et al., 2007; Rosilawati et al., 2014).

CHANGES IN SOIL UNDER MOIST CONDITION

Under moist condition, the pH is increased from 3.43 to 4.02 due to application of 6 tonnes GML/ha in 2 weeks and exchangeable Al is decreased about 3 $cmol_c$/kg soil in the 12th week. For rice production, this level is considered too high. Hydrated lime about 6 tonnes/ha applied at third week maintains pH at 5. Liquid lime shows similar trend as submerged causing no effect on pH but a decrease in Al.

EFFECT OF GROUND MAGNESIUM LIMESTONE IN SOIL MANAGEMENT

Acid sulfate soil, high in acidity, and Al^{3+} treated with GML for its management works effectively by precipitating Al^{3+} as $Al(OH)_3$. GML releases OH^- ion from its carbonate reaction, which reacts to form inert precipitates of aluminum ions decreasing Al concentration in the soil.

$$(Ca, Mg)(CO_3)_2 \rightarrow Ca^{2+} + Mg^{2+} + CO_3^{2-}$$

$$CO_3^{2-} + H_2O \rightarrow HCO_3^{-1} + OH^{-1}$$

$$Al^{3+} + 3OH^{-1} \rightarrow Al(OH)_3$$

Effect of Hydrated Lime in Soil

Calcium hydroxide, called hydrated lime, quick lime, slaked lime and milk of lime, when applied on acid sulfate soil releases Ca^{2+} and OH^{-1}. OH^{-1} ameliorates the soil by precipitating Al and by increasing the pH.

Effect of Liquid Lime in Soil

It provides calcium and magnesium to the soil with quick effect on soil acidity. Calcium and magnesium ameliorate the infertility of the soil; recommended doze is 20 L/ha. It is easily absorbed and filters in the soil but quickly evaporates in the atmosphere.

Effect of Fused Magnesium Phosphate in Soil

FMP is applied as a subsidy by the Malaysian government on the acid sulfate soils in Kelantan with recommended rate of 125 kg/ha. It is proved not as effective as GML. FMP is reacted in the soil accordingly:

$$CaMgP_2O_7 + H_2O \rightarrow Ca^{2+} + Mg^{2+} + 2HPO_4^{2-}$$

The active ingredients of fused magnesium phosphate are P_2O_5 (15%), CaO (28%), MgO (15%), SiO_2 (24%), and Fe (3%). It is good to apply on paddy rice fields.

Effect of Burnt, Partial-Burnt, and Zero-Burnt Techniques

The data were summarized to coincide approximately with the three different practices of land preparation for oil palm planting, that is, burning, which was generally the case in 1982 and earlier, partial burning (1983–1993), and the zero-burn technique which is the most common method from 1994 onward. Only the burnt and zero-burnt periods will be discussed in order to avoid complication in the interpretation of the partial-burnt period, which might include some fields that were fully burnt, whereas others were zero burnt.

Broadcasting of the burned biomass of the oil palm at tree planting did not show long-lasting effects on the soil pH, whereas rise in soil pH for short span had been testified by Ling et al. (1979), which was ascribed to the discharge of exchangeable bases in the ash. This can be implied from these two outcomes that the ash imposed impermanent influence on soil pH and simply abolished by the utilization of acidifying fertilizers to the palms. Practicing of zero-burnt technique intensified the soil organic C content in the inter-rows (IR) and in both soil layers as shown in Table 7.4.

Generally, all the main soil nutrients of total N, total P, Bray-2P, and exchangeable Mg mount up during the zero-burnt phases. Conversely, concentration of soil

TABLE 7.4
Percentage Change for Each Soil Chemical Parameter for Different Periods

Parameter/ Variable		Change (%) of Each Parameter at Two Sites and Two Depths During Burnt, Partial-Burnt, and Zero-Burnt Period[a]			
		Site		Depth	
		IR	P. Circle	0–15 cm	15–45 cm
pH	Burnt	−3.49	−2.35	−2.41	−3.21
	Partial Burnt	−3.63	−1.04	−1.61	−2.49
	Zero burnt	−1.75	−2.24	−2.95	−1.59
Organic C (%)	Burnt	−9.63	–	−5.49	−11.67
	Partial burnt	−18.97	–	−12.43	−24.57
	Zero burnt	7.77	–	1.87	15.56
Total N (%)	Burnt	−2.38	–	−1.63	−3.88
	Partial burnt	−14.31	–	−8.08	−19.35
	Zero burnt	7.03	–	0.00	14.29
Total P (mg/Kg)	Burnt	−14.37	−6.72	−6.10	−16.94
	Partial burnt	9.37	0.00	17.88	−3.28
	Zero burnt	31.04	−4.26	−12.47	7.80
Bray-2P (mg/Kg)	Burnt	−4.37	−20.82	−3.45	−17.71
	Partial burnt	30.07	−1.67	20.27	−2.30
	Zero burnt	38.42	0.89	12.00	72.04
Exch. K (cmol/Kg)	Burnt	18.99	0.00	9.72	13.25
	Partial burnt	8.33	−5.90	0.00	−2.94
	Zero burnt	1.09	27.71	5.89	9.17
Exch. Mg (cmol/Kg)	Burnt	−23.30	8.25	0.00	−16.25
	Partial burnt	−26.12	−30.71	−26.42	−31.58
	Zero burnt	58.46	22.79	23.33	58.46

[a] Values quoted are the mean values.

exchangeable K was maximum when burning was practiced. This was amenable with the clarifications made by Sly and Tinker (1962) that burning suppressed nitrogen and dropped magnesium and calcium in the soil but provided higher exchangeable K. The upsurge of total N during the zero-burnt phase paralleled with the accumulation in soil organic C, which suggested that preceding palm biomass had returned a large content of N to the soil. In contrast, highly frequent application of P and K fertilizer to the palm resulted in the considerable increase in Bray 2-P and exchangeable K in the palm circle with subsequent excellent yield obtained from these nutrients (Goh and Kee, 2000). This study demonstrates that the adaptation of sound fertilizer management practices and zero-burnt replanting strategy has boosted the soil fertility status of the Ultisols under oil palm.

AGRONOMIC PROBLEMS

Agronomic problems encountered in acid sulfate soils are extremely difficult to alleviate. Crops are facing with toxic amounts of Al and Fe in these soils as their concentrations are too high at low pH. Phosphate availability is decreased by Fe–Al–phosphate interaction, leading to P deficiency. Nutrient deficiency, toxicity due to Fe(II) and hydrogen sulfide, and physical problems arisen from inhibition of root development in sulfuric horizon with low pH and high Al contents are some other problems for plant growth in acid sulfate soils (Dent, 1986).

The CEC of typical acid sulfate soils in Malaysia is moderately high, with values higher than the normal soils in the country such as the Oxisols and Ultisols that are found extensively in the upland areas. This moderately high value of the CEC for acid sulfate soils is related to the soil mineralogy. The clay fraction of acid sulfate soils is dominated by kaolinite but with a fair amount of smectite and vermiculite (Shamshuddin et al., 1986; Shamshuddin and Auxtero, 1991). Smectite and vermiculite by themselves are known to have high CEC.

Some of the acid sulfate soils in Peninsular Malaysia are planted with cocoa. Al present in acid sulfate soils affects the quality of cocoa bean, resulting in a lower price at the marketplace. Water management is the key to successful implementation of agricultural enterprise on acid sulfate soils in view of the need for adequate drainage. Of even more importance is the danger posed by developing acidity if the soils are overdrained. This is associated with the release of high amounts of Al into the soils, which is toxic to cocoa plants. Liming is not a viable agronomic practice to ameliorate acid sulfate soils because of the high amounts of lime (>10 tonnes/ha) required to neutralize the soil acidity. Organic matter application can eliminate Al toxicity in acid sulfate soils through formation of Al–organic complex. This interaction helps account for a better plant growth at low pH values. Peat, chicken dung, POME, rice straw, and green manure are some kinds of organic materials, which are abundant in Malaysia.

Oil palm is severely affected by the presence of high amounts of Al in the soil on which it grows. Aluminum affects the growth of the roots of the oil palm. Oil palm thrives well under moist conditions; hence, sandy to silty clay textures are the best soil textures for retaining adequate amounts of water required by oil palm. Flooding acid sulfate soils is found to increase oil palm yield from 5.7 to 17.5 FFB/ha/year (Poon, 1977). It is therefore advisable not to drain the soil during land preparation for oil palm cultivation. Drains are only dug to remove the excess water present in the area. This method has been proven to be a viable agronomic practice. The water table level should be always above the pyritic layer. We should never allow the water table level to go down below the pyritic layer. If that happens, pyrite undergoes oxidation, thereby releasing acidity that causes problems for the growing oil palm on the acid sulfate soils. Phosphate rock has to be applied regularly. Phosphate is able to prevent Al release by virtue of it being able to react with Fe to form insoluble iron phosphate that coats pyrite crystal. Limestone and bunch ash are also used for amelioration of acid sulfate soils for oil palm production to increase the pH. Bunch ash is produced by incineration of thrashed oil palm fruit bunch. It is extremely basic (pH 12), containing 41% K_2O, 4% P_2O_5, 6% MgO, and 5% CaO. Bunch ash increases K and limestone increases Ca and Mg for oil palm growth, and their combination can increase the oil palm yield considerably in acid sulfate soils.

Rice fields in acid sulfate soils are contaminated with Fe oxides/hydroxides. Fe(II) may be released in toxic amounts in paddy fields. The water is reddish in color, showing the presence of high amounts of soluble iron. Its concentration reaches to 5000 mg/kg within 2 weeks of flooding, whereas values above 500 mg/kg are toxic to rice (Ponnamperuma et al., 1973). Al toxicity is also one of the major limitations for rice cultivation in acid sulfate soils. Liming with GML is a viable practice to ameliorate acid sulfate soils for rice cultivation. Ca and Mg are able to reduce the toxic effect of Al. Another major problem for paddy in acid sulfate soils is phosphorus deficiency. The required P in soil for rice production is 7–20 mg/kg (Dobermann and Fairhurst, 2000). When soluble phosphate fertilizer is applied, it will be fixed by Al and Fe and consequently converted to insoluble form. Fused magnesium phosphate application can alleviate this problem. Adding organic fertilizer into a flooded acid sulfate soil would intensify reducing condition, resulting in release of Fe^{2+}, which is toxic to rice plants (Tran and Vo, 2004).

THE ROLE OF LEGUME COVERS IN SOIL MANAGEMENT IN MALAYSIA

Table 7.5 shows the characteristics of the legumes planted as ground covers in Malaysia, which are described by Chiu et al. 2011. Upon achieving 100% ground coverage (approximately 8 months after planting), the eroded topsoils were reduced from 7.5 tonnes/ha of soils to almost negligible (Arif et al., 2007) as given in Figure 7.2.

TABLE 7.5
Characteristics of Common Leguminous Cover Plants Planted in Malaysia

Species	Plant Characteristics	Nitrogen (kg/ha) Accumulated	Fixed	%	Seeding Rate (kg/ha)
Pueraria javanica	Vigorous, thick, and dominant without overhead shade, suppresses other undesirable plants	13–225	6–172	44–76	2.0–4.0
Calopogonium caeruleum	Slow initially, spreads vigorously later, shade tolerant	–	–	–	0.5–1.0
Calopogonium mucunoides	Growth rapid initially, but dies early, unpalatable to cattle	4–108	1–76	35–79	4.0–6.0
Centrosema pubescens	Fast establishing, very susceptible to pests	6–132	3–104	48–79	4.0–6.0
Mucuna cochinchinensis	Rapid coverage, but short lived (6–9 months) and drought sensitive	22–305	6–224	36–86	4.0–8.0
Mucuna bracteata	Very vigorous, thick, and dominant, suppresses other weeds, shade and drought tolerant. Unpalatable to cattle when mature	180–522	126–365	33–88	0.1–0.2

Source: Modified from Chiu, S.B. et al., Agronomic principles and practices of groundcover management of oil palm, in *Seminar on Agronomic Principles and Practices of Oil Palm Cultivation*, Agricultural Crop Trust, Sibu, Malaysia, pp. 199–231, 2011.

FIGURE 7.2   Good legume (*Mucuna bracteata*) cover plants in an immature oil palm fields at 8 months after legume planting. (From Ng, P.H.C. et al., The use of phosphate rocks for growing Mucuna bracteata in oil palm legume systems to enhance sustainability, in *Proceedings of MSSS 2010 International Conference*, April 12–16, 2010, Kuantan, Malaysia, 2010.)

Runoff losses were reduced by 30% of rainfall in areas covered by legume compared to bare ground (Ng et al., 2008). Another major benefit of leguminous cover is the nitrogen fixation and incorporation of the fixed nitrogen into the soils. About 250 kg N/ha was fixed by legumes in addition to other immobilized nutrients (Ng et al., 2010). Better availability of applied PRs (Zaharah and Bah, 1997), improved soil structures, and probably an increased quantity, diversity, and activity of soil microfauna, and weed controls (Chiu et al., 2011) due to the deposition of dead leaf residues on top of the soil are the other benefits of planting legume cover crops as presented in Figure 7.3.

FIGURE 7.3   Leaf litters formed underneath the legume cover plant. (From Chiu, S.B. et al., Agronomic principles and practices of groundcover management of oil palm, in *Seminar on Agronomic Principles and Practices of Oil Palm Cultivation*, Agricultural Crop Trust, Sibu, Malaysia, pp. 199–231, 2011.)

REFERENCES

Anda, M., Shamshuddin, J., and Fauziah, C.I. (2013). Increasing negative charge and nutrient contents of a highly weathered soil using basalt and rice husk to promote cocoa growth under field conditions. *Soil and Tillage Research*, 132: 1–11.

Anda, M., Shamshuddin, J., and Fauziah, C.I. (2015). Improving chemical properties of a highly weathered soil using finely ground basalt rocks. *Catena*, 124: 147–161.

Arif, S., Shahrakbah, Y., and Kee, K.K. (2007). Impact of leguminous covers and palm chips on soil nutrient losses in oil palm replants. In: *Proceedings of International Conference on Oil Palm and Environment (ICOPE)*, Bali, Indonesia, 16 p.

Auxtero, E.A. and Shamshuddin, J. (1991). Growth of oil palm (Elaeis guineensis) seedlings on acid sulfate soils as affected by water regime and aluminium. *Plant and Soil*, 137(2): 243–257.

Bah, A.R. and Zaharah, A.R. (2001). Gliricidia (Gliricidia sepium) green manures as a potential source for maize production in the tropics. In: *Optimizing Nitrogen Management in Food and Energy Production and Environmental Protection: Proceedings of the 2nd International Nitrogen Conference on Science and Policy*, The Scientific World 1, CRC Press, Potomac, MD.

Caliman, J.P., Hardianto, J., and Ng, M. (2001). Strategy for fertilizer management during low commodity price. In: *Cutting-edge technologies for sustained competitiveness: Proceedings of the 2001 PIPOC International Palm Oil Congress*. Agriculture Conference, Malaysian Palm Oil Board (MPOB), Kuala Lumpur, Malaysia, August 20–22, 2001, pp. 295–312.

Che Johari, M. (2005). Malaysian palm oil board sedia gerakkan pekebun capai visi 35:25. *Berita Sawit*, 43: 3.

Chew, P.S., Kee, K.K., Goh, K.J., Quah, Y.T., and Tey, S.H. (1992). Fertilizer management in oil palms. In: B. Aziz et al., (eds.) *Fertilizer Usage in the Tropics*. Malaysian Society of Soil Science, Kuala Lumpur, Malaysia, pp. 43–64.

Chintu, R. and Zaharah, A.R. (2003). Nitrogen uptake of maize (Zea mays, L.) from isotope-labeled biomass of Paraserianthes falcataria grown under controlled conditions. *Agroforestry Systems*, 57(2): 101–107.

Chiu, S.B., Gatot, A.R., and Saiful, R.A. (2011). Agronomic principles and practices of groundcover management of oil palm. In: Goh, K.J., S.B., Chiu, and S. Paramananthan, (Eds.) *Seminar on Agronomic Principles and Practices of Oil Palm Cultivation*. Agricultural Crop Trust, Sibu, Malaysia, pp. 199–231.

Corley, R.H.V. and Tinker, P.B.H. (2008). The oil palm. Publisher: John Wiley and Sons.

Dent, D. (1986). *Acid Sulphate Soils: A Baseline for Research and Development*. ILRI Publication 39, International Institute for Land Reclamation and Improvement, Wageningen, The Netherlands.

Dobermann, A. and Fairhurst, T. (2000). *Rice: Nutrient disorders & nutrient management* (Vol. 1). International Rice Research Institute, Los Banos, Phosphate & Potash Inst (PPI) & Potash & Phosphate Institute, Canada.

Duff, D. (1993). *Holmes' Principles of Physical Geology*. Chapman and Hall, London.

Fageria, N.K. and Baligar, V.C. (2008). Ameliorating soil acidity of tropical Oxisols by liming for sustainable crop production. *Advances in Agronomy*, 99: 345–399.

Gillman, G.P. (1985). Influence of organic matter and phosphate content on the point of zero charge of variable charge components in oxidic soils. *Soil Research*, 23(4): 643–646.

Goh, K.J. and Härdter, R. (2003). General oil palm nutrition. In: Fairhurst, T.H. and R. Härdter, (Eds.) *Oil palm: Management for Large and Sustainable Yields*, PPI/PPIC-IPI, Singapore, pp. 191–230.

Goh, K.J. and Kee, K.K. (2000). The oil palm sector in Southeast Asia: Changing requirements for fertilizers, particularly P and K. In: Workshop on Improving Soil Fertility Management in Southeast Asia, November 21–23, 2000, Bogor, Indonesia. IBSRAM, Thailand.

Hardy, F. (1971). Soil conditions and plant growth. *Cocoa Grower's Bulletin*, 17: 27–30.

Hue, N.V. (1992). Correcting soil acidity of a highly weathered ultisol with chicken manure and sewage sludge. *Communications in Soil Science and Plant Analysis*, 23(3–4): 241–264.

Hue, N.V. and Amien, I. (1989). Aluminum detoxification with green manures 1. *Communications in Soil Science and Plant Analysis*, 20(15–16): 1499–1511.

Hue, N.V., Craddock, G.R., and Adams, F. (1986). Effect of organic acids on aluminum toxicity in subsoils. *Soil Science Society of America Journal*, 50(1): 28–34.

IBSRAM. (1985). *Report of the Inaugural Workshop and Proposal for Implementation of the Acid Tropical Soils Management Network*. IBSRAM, Bangkok.

Kanapathy, K. (1973). Reclamation and improvement of acid sulphate soils in West Malaysia. In: Dost, H., (Ed.) *Acid sulphate soils*. International Institute for Land Reclamation and Improvement, Wageningen, The Netherlands, Vol. 1, pp. 383–390.

Kimi, S. and Daham, M.D. (1993). Managing salinity problems in rice fields. In *Proceedings of the International Conference for Agricultural Machinery and Process Engineering*, pp. 556–564. Seoul, Korea.

Lim, K.C. and Zaharah, A.R. (2002). The effects of empty oil palm fruit bunches on oil pal nutrition and yield, and soil chemical properties. *Journal of Oil Palm Research*, 14(2): 1–9.

Ling, A.H., Tan, K.Y., Tan, P.Y., and Sofi, S. (1979). Preliminary observation on some possible post clearing changes in soil properties. Seminar on Fertilizer and Management of deforested land. Kota Kinabalu, Sabah.

MARDI. (1990). *Panduan Menanam dan Pemprosesan Koko*. Berita Publishing Press, Kuala Lumpur, Malaysia.

MARDI. (2001). *Composting the Waste from a Rice Processing Plant*. Malaysian Agricultural Research and Development Institute, Serdang, Malaysia.

Md. Mahmudul, A., Chamhuri, S., Md. Wahid, M., Rafiqul, I.M., and Mohd. Ekhwan, T. (2010). Socioeconomic profile of farmers in Malaysia: Study on integrated agricultural development area in North-West Selangor. *Agricultural Economics and Rural Development*, 7(2): 249–265.

Miller, R.W. and Donahue, R.L. (1990). *Soils: An introduction to soils and plant growth* (6th ed.). Prentice-Hall International, Englewood Cliffs, NJ.

Mokhatar, S.J. and Daud, N.W. (2011). Performance of *Hevea brasiliensis* on Haplic Acrisol soil as affected by different source of fertilizer. *International Journal of Applied*, 1(1).

Ng, P.H.C., Goh, K.J., Gan, H.H., Yacob, S., Kee, K.K., and Teo, C.B. (2008). Introduction of shade-tolerant cover crop, Mucuna bracteata, enhances the sustainability of oil palm (Elaeis guineensis). In: *Proceedings of 5th International Crop Science Congress 2008*, April 13–18, 2008, Jeju, South Korea.

Ng, P.H.C., Goh, K.J., Gan, H.H., Yacob, S., and Zaharah, A.R. (2010). The use of phosphate rocks for growing Mucuna bracteata in oil palm legume systems to enhance sustainability. In: *Proceedings of MSSS 2010 International Conference*, April 12–16, 2010, Kuantan, Malaysia.

Ooi, L.H. (2011). Agronomic principles and practices of replanting oil palm. In: Goh, K.J., S.B. Chiu, and S. Paramananthan, (Eds.) *Seminar on Agronomic Principles and Practices of Oil Palm Cultivation*, Agricultural Crop Trust, Sibu, Malaysia, pp. 171–197.

Othman, A.S. (1993). *Pengeluaran Koko di Malaysia*. Dewan Bahasa dan Pustaka, Kuala Lumpur, Malaysia.

Paramananthan, S. (2000). *Soils of Malaysia: Their Characteristics and Identification*, Vol. 1. Academy of Sciences Malaysia, Kuala Lumpur, Malaysia.

Ponnamperuma, F.N., Attanandana, T., and Beye, G. (1973). Amelioration of three acid sulphate soils for lowland rice. In *Acid sulphate soils. Proceedings of the International Symposium, Wageningen, ILRI Publ*, Vol. 18, pp. 391–406.

Poon, Y.C. (1977). The amelioration of acid sulfate soils with respect to oil palm. *Tropical Agriculture*, 54(4): 289–305.

Rosilawati, A.K., Shamshuddin, J., and Fauziah, C.I. (2014). Effects of incubating an acid sulfate soil treated with various liming materials under submerged and moist conditions on pH, Al and Fe. Vol. 9(1), pp. 94–112, January 2, 2014. DOI: 10.5897/AJAR12.289 ISSN 1991–637X ©2014 Academic Journals http://www.academicjournals.org/ AJAR.

Seefeldt, S. (2013). Animal manure as fertilizer. University of Alaska Fairbank, Cooperative Extension Service: www.uaf.edu/ces or 1-877-520-5211.

Shamshuddin, J., Mokhtar, N., and Kamal, A.J. (1985). Chemical and mineralogical characteristics of an organic soil (Troposaprist) from Sg. Burong, Tg. Karang, Selangor. *Pertanika*, 8(3): 299–304.

Shamshuddin, J. and Auxtero, E.A. (1991). Soil solution compositions and mineralogy of some active acid sulfate soils in Malaysia as affected by laboratory incubation with lime. *Soil Science*, 152(5): 365–376.

Shamshuddin, J., Paramananthan, S., Wan, N., and Mokhtar, N. (1986). Mineralogy and surface charge properties of two acid sulfate soils from Peninsular Malaysia. *Pertanika*, 9(2): 167–176.

Shamshuddin, J., Sharifuddin, H.A.H., and Bell, L.C. (1998). Changes in chemical properties of an ultisol as affected by palm oil mill effluent application. *Communications in Soil Science and Plant Analysis*, 29(15–16): 2395–2406.

Sharifuddin, H.A.H., Yusuf, M.N.M., Shamshuddin, J., and Norhayati, M. (1989). Management of soil acidity in Malaysia, *ACIAR Monograph*, 13: 118 p.

Singh, K., Ishii, T., Parco, A., Huang, N., Brar, D.S., and Khush, G.S. (1996). Centromere mapping and orientation of the molecular linkage map of rice (Oryza sativa L.). *Proceedings of the National Academy of Sciences*, 93(12): 6163–6168.

Sly, J.M.A. and Tinker, P.B. (1962). An assessment of burning in the establishment of oil palm plantations in southern Nigeria. *Tropical Agricultural Trinidad*, 39: 271–280.

Soil Survey Staff. (2010). *Keys to Soil Taxonomy*. United States Department of Agriculture, Washington DC.

Stevenson, F.J. (1994). *Humus Chemistry: Genesis, Composition, Reactions*. John Wiley & Sons, New York.

Suswanto, T., Shamshuddin, J., Omar, S.S.R., Mat, P., and Teh, C.B.S. (2007). Alleviating an acid sulphate soil cultivated to rice (*Oryzasativa*) using ground magnesium limestone and organic fertilizer. *Jurnal Tanah danLingkungan*, 9(1): 1–9.

Tan, K.H. and Binger, A. (1986). Effect of humic acid on aluminum toxicity in corn plants. *Soil Science*, 141(1): 20–25.

Tessens, E. and Shamshuddin, J. (1983). *Quantitative Relationships Between Mineralogy and Properties of Tropical Soils*. Universiti Pertanian Malaysia, Selangor, Malaysia.

Tessens, E. and Zaharah, A.R. (1983). The residual influence of P-fertilizer application on soil pH values. *Pedologie*, 32: 367–368.

Thong, K.C. and Ng, W.L. (1980). Growth and nutrients composition of monocrop cocoa plants on inland Malaysian soils. In: *Proceedings International conference on Cocoa and Coconuts, 1978*. Incorporated Society of Planters, Kuala Lumpur, Malaysia, pp. 262–286.

Tie, Y.L. (1982). Soil classification in sarawak. Technical Paper No. 6, Department of Agriculture, Sarawak, Malaysia.

Tran, K.T. and Vo, T.G. (2004, August). Effects of mixed organic and inorganic fertilizers on rice yield and soil chemistry of the 8th crop on heavy acid sulfate soil (Hydraquentic Sulfaquepts) in the Mekong Delta of Vietnam. In: *A paper presented at the 6th International Symposium on Plant-Soil at Low pH*, Sendai, Japan.

Yacob, O. and Sulaiman, W.H.W. *The Management of Soils and Fertilizers for Sustainable Crop Production in Malaysia*. Faculty of Agriculture, University Pertanian Malaysia. Serdang, Malaysia.

Yew, F.K. (2000). Impact of zero burning on biomass and nutrient turnover in rubber replanting. In *International Symposium on Sustainable Land Management*, Sri Kembangan, Kuala Lumpur.

Zaharah, A.R. and Bah, A.R. (1997). Effect of green manures on P solubilization and uptake from phosphate rocks. *Nutrient Cycling in Agroecosystems*, 48(3): 247–255.

Zaharah, A.R., Hawa, J., and Sharifuddin, H.A.H. (1985). Accumulation and migration of phosphate applied as rock phosphate in an oil palm plantation. *Pertanika*, 8(3): 317–321.

8 Future Soil Issues

Roslan Bin Ismail and Nur Zurairahetty Mohd Yunus

CONTENTS

INTRODUCTION

Soils in Malaysia are limited natural resource. In the last 30 years, Malaysia has rapidly changed from agriculture as the main source of income to industrialization, petroleum, manufacturing, tourism, and many others. This gave the country a well-projected steady growth of 7%–8% of yearly gross domestic product (GDP). With such an economic growth, land became a primary commodity for the country. Agriculture was facing a new challenge from industrialization of Malaysia. Now, land has become a very limited source, therefore utilization of idle/abandoned land becomes a viable option. In 1990, a study conducted by Aminuddin et al. (1990) shows land use for agricultural purposes as shown in Table 8.1.

Malaysia structural transformation since independence has had a major impact especially on the agriculture sector. The GDP from agriculture shows a steady decline from 51.5% in 1965 to 20.5% in 1995 and further reached projected level of 15.5% in year 2000. The manufacturing sector on the other hand had risen vice versa up to 37.5% in year 2000 taking a heavy toll on the land/soil availability for crop production. This is a dilemma of agriculture, competing with industrialization for a piece of land.

Malaysia is classified as Af in Köppen climate classification systems: it indicates the general hot and humid climatic condition throughout the year. Thus, the soil weathering rate is high, giving rise to land areas with highly weathered soils. These types of soils are normally formed through in situ weathering of igneous and high-grade metamorphic rocks as well as sedimentary and low-grade metamorphic rocks known as sedentary soils. These sedimentary soils with well-developed soil horizons often vary in soil horizon/depth in addition to medium to high soil acidity and with low to moderate soil fertility status. In general, it occupies the central part of Peninsular Malaysia and the interiors of Sarawak and Sabah. Following that, there are also alluvial soils (old, recent, and subrecent); soils landscapes with undulating and rolling topography can be observed as we move further to the coastline. The soils in the coastline are often of low fertility with poor drainage that reflects the young soils profile development, inclusive of peat soils.

However, some variations are available in distinction of upland soils between Peninsular Malaysia, Sarawak, and Sabah. Uplands are those areas where the soils

TABLE 8.1
Distribution of Various Categories of Land in Malaysia

Region	Peninsular Malaysia	Sarawak	Sabah
Total land area (million ha)	13.16	12.24	7.63
Land area suitable for agriculture (million ha)	6.19	1.81	2.31
Land area unsuitable for agriculture (million ha)	6.85	10.43	5.25
Percentage of unsuitable land (%)	52	85	69
Upland areas unsuitable for agriculture (million ha)	5.61	8.57	4.49
Unsuitable upland as a percentage of total unsuitable land (%)	82	82	87
Steep land (% of total land)	36	70	22

Source: Aminuddin, B.Y., Resources and problems associated with sustainable development of upland areas in Malaysia, in *Technologies for Sustainable Agriculture on Marginal Uplands in Southeast Asia,* eds. G. Blair and R. Lefroy, ACIAR Proceedings no. 33, Australian Centre for International Agricultural Research, Canberra, pp. 55–61, 1990.

are sedentary or have developed from old alluvium. Steep lands are sloping lands defined on the basis of slope gradient (Hashim, 2009). In Peninsular Malaysia, steep lands are lands with gradients >20° (±36%), whereas in Sarawak and Sabah (also known as Borneo), the corresponding slope gradients are divided into two sections: (1) slope with >33° (±65% in average) and (2) slope with >25° (±47% in average). This consensus was made due to hill and mountainous geographic condition of the island of Borneo.

Generally, steep lands are considered as areas where the native forest cover should be maintained and not disturbed (DOA-MARDI, 1993; Siong, 2003). This rule is imposed because of the vulnerability of such land to the processes of soil erosion. However, the rule is not always adhered to. Indeed, soil erosion is one of the major forms of land degradation recognized in Malaysia (CAP, 2000; DOA, 2013). Besides that, *problem soils* are another soil issues that need to be addressed in Malaysia. The idle/abandoned land is often problem soils. This chapter aims to look at the future soil issues from the perspective of idle/abandoned land in Malaysia. By understanding the nature, extent, and issues related to these types of soils, it is expected that good soil and land management can be developed for future sustainable land use.

Soils in Malaysia are divided into three geographical regions, namely, Peninsular Malaysia, Sarawak, and Sabah. Each of the regions has their distinctive geographic condition as stated earlier. In Peninsular Malaysia, coastal plains with relatively fertile alluvial soils are found on the west coast. Alluvial soils are also present in some parts of the east coast. In the uplands, the dominant soils are sedentary soils that are strongly weathered. There are also several groups of soils, which are normally referred to as *problem soils*. They are referred as problem soils due to some physical and chemical (localized) constraints to plant growth that these soils exhibit. If not for the constraints, these soils can be utilized for crop growth (food production). Many of these problematic soils are located near the water source and provide ease of access to road transportation, thus giving an advantage in location and topography for farming activity. Examples of these soils are as follows:

1. Sandy beach soils or locally known as *tanah beach ridges interspersed with swales* (BRIS): distinctive of inherent low soil fertility status to moderate (require some soil amelioration practices) and low water-holding capacity due to sandy nature of these soils. Land size of these soils is estimated to be about 154,400 ha; these soils are found mainly in the east coast of Peninsular Malaysia.
2. Acid sulfate and potential acid sulfate soils: land size is about 467,800 ha; these soils are wide spread in the east and west coast.
3. Peat soils with an area of 768,200 ha: widespread in southern Peninsular Malaysia and Sarawak, with some also recorded in Sabah.
4. Wasteland created by tin-mining activities, known as *tin tailings*, with an area of 200,000 ha is found mainly in the west coast.

Generally, these four groups are often referred as idle land in Malaysia. These are often underutilized and/or left abandoned due to low economic value. The problem of idle land became so apparent that in 1980s, the Ministry of Agriculture ordered a special study to be conducted by setting up a Task Force for idle land. The aims to identify nationwide idle plots and their possible remedies are put forward. Since then, various actions have been taken by Ministry of Agriculture with the Department of Agriculture as the executor of the remedial plans. Besides that, Rural Development and Regional Development and private agencies also had their effort concentrated on the utilization of the idle land. The aim is to induce farmers and land owners to continuously utilize their lands, hence minimizing idle land scenario.

For example, in the case of rice farming, two major initiatives (short named as Kemubu Agricultural Development Authority [KADA] and Muda Agricultural Development Authority [MADA]) in land rehabilitation were taken by the Department of Agriculture dating back to 1980s. To date, some positive effect in yield return to investment is seen. In other crop sectors, Federal Land Development Authority (FELDA), Federal Land Consolidation and Rehabilitation Authority (FELCRA), South Kelantan Development Authority (KESEDAR) focus on oil palm, meanwhile, RISDA (rubber), MCB (cocoa), the Ministry of Agriculture, and the Department of Agriculture Sabah, with mixed crop, have made significant progress in their effort to consolidate and rehabilitate the soil; thus, idle land is very well minimized. Although some efforts were successful, many were short lived for various reasons.

There is no standard definition available for idle and/or abandoned land. Different agencies have different outlook on the idle land issues. However, the Ministry of Agriculture Task Force for Idle Land defines idle land as the following:

1. Agricultural crop land to which farmers have been granted freehold titles but which has not been cultivated or utilized for grazing purposes for three consecutive years.
2. Parcels of land to which farmers have been granted freehold titles and gazetted for cultivation; or temporary occupancy of land (TOL) pending the

issuance of permanent titles, but which have not been in any way cultivated in compliance with the stipulated conditions.
3. Land equipped with the necessary physical infrastructure for rice double-cropping in a year but instead is being utilized for single-cropping.

Not included in the definition are areas of noncultivated government land and forest land belonging to the government, which have been gazetted for agricultural purposes but are still undergoing clearing and cleaning up.

On the other hand, FELCRA definition of idle land is better defined as it covers ex-mining land, sandy soils (tanah BRIS), as well as forest land which has been logged and not managed for forest plantation.

Until now, the exact extent of the idle land remains unsolved. No one knows exactly how much land throughout the country has been neglected or left idle. Back in 1978, a survey conducted by the Prime Minister's Department estimated that about 565,000 ha was marked in West Malaysia that comprised all types of agricultural crop land, and then in 1981, the Task Force for Idle Land estimated the total area of about 890,000 ha (West Malaysia) idle land that separated between rice land and nonrice land as presented in Table 8.2. This shows an increment of 35% of idle land recorded within 3 years of survey. The land size is believed to be more than 1.0 million ha of idle land; unfortunately, no data were made available for East Malaysia (Sabah and Sarawak).

Idle and/or abandoned land belongs to two categories in Malaysia: permanent and temporary abandonment as defined in the following:

1. Permanent abandonment, also known as continuous abandonment, refers to land, which has been abandoned for a long duration whereby rehabilitation efforts would be very costly.
2. Temporary abandonment refers to land that is cultivated during the main season and left idle during the off season due to irrigation, pest/disease, and/or related problem, that is, rice farming.

The causes of idle land are varied and are mostly associated with physical, economic, and social or institutional factors. The summary of these factors is shown in Figure 8.1. The individual factors involved in idle/abandoned land have been highlighted by many authors (Ely and Wehrwein, 1940; Huang, 1972; Sahak, 1986; Smith and Nasr, 1992; Azima and Ismail, 2011).

TABLE 8.2

West Malaysia: Area of Abandoned Land (in hectares) by Agencies and Categories, 1981

Agency	Rice Land	Nonrice Land	Total (ha)
Ministry of Agriculture	160,863	–	160,863
FELCRA	–	565,140	565,140
Other agencies	–	163,997	163,997
Total	160,863	729,137	890,000

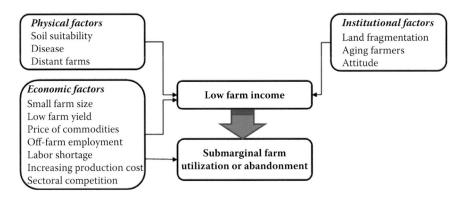

FIGURE 8.1 Factors influencing land abandonment. (From Consumer Association Penang [CAP], *Land Issues in Malaysia*, Consumer' Association of Penang [CAP], Penang, Malaysia, 2000.)

From Figure 8.1, it is clear that soil suitability and crop disease are physical factors. Both have remained as a major problem facing agriculture sector, combined with severe weather conditions such as flash flood and drought. Rice blast and *brown planthopper disease* for instance, occurred in 1983, destroyed almost entire rice crop in MUDA project area in one season. In the subsequent year, rice yield was reduced drastically. The recovery from that faithful year is still ongoing. The disease is still there but minimal with some control and precaution measurement before widespread attack occurs again. Thus a policy was laid out known as National Agricultural Policy (NAP).

NAP was introduced in 1984. The aim is to achieve 80% self-sufficiency in major food products, especially rice. Vegetables, fish, poultry, and eggs have reached the target of self-sufficiency; domestic rice production has to reach that level yet. Some decline was observed in beef (30%), mutton (10%), milk (4%) and rice self-sufficiency from 73% (1990) to 62% (1995) and increased again to 72% in the year 2008. The numbers have to cross 80% mark. Based on the current situation, existence of underutilized idle land would only limit the possibility of achieving the self-sufficiency target of Malaysia.

As stated earlier, sandy soils, acid sulfate soils, and wasteland are mainly from mineral soils. A good easy read on acid sulfate soils in Malaysia is available, which is written by Shamsuddin (2006), expert on the subject matter. Peat soils are organic soils. Peats are formed by the accumulation of organic soil materials. These materials consist of undecomposed, partially decomposed, and highly decomposed plant remains, whereby soil scientist refer them as saprist, fibrist, and hemist, respectively. Peats in Malaysia are classified as tropical lowland peats; they form a fragile ecosystem because they are dome shaped and are almost 100% organic. They are able to store large quantity of water, minimize flooding hazard, and retain contaminants such as heavy metals.

In the last 15 years, due to population pressure, the need to produce more food and to eradicate rural poverty, the government has drained and developed peatlands for agriculture. It started when Gurmit et al. (1986) introduced water control and nutritional management of oil palm cultivation on peat. It created a wave of peat-lands that are cleared and developed for oil palm. However, it created environmental issues such as release of carbon dioxide (CO_2) and nitrous oxide (N_2O) that contribute to global warming; thus, success of peatland usage became limited with a dilemma to conserve or to develop tropical lowland peatland. Malaysia is estimated to have about 2.5 million ha of peatland with southern state in Peninsular Malaysia; Johor has the highest peatlands at 205,856 ha. Generally, pineapple is cultivated in Johor, indicating potential of peatlands as agricultural soil. As such, leaving such an extent land size idle/underutilized is a waste of land resource. No policy or guideline is made available to manage these peatlands; thus, Paramananthan (2014) stated that Malaysia needs a National Peatland Policy.

Besides soil types, issues, and their use for future development, a brief look at some other factors is also worthy of consideration in planning national land use for agriculture and other development. Factors such as rainfall characteristics, soil erosion and land degradation monitoring (Hashim, 2009), and native people land status are also important.

According to Hashim (2009), land degradation needs to be monitored regularly as it may have negative effects on agricultural production if it occurs in agricultural land. Malaysian Centre for Remote Sensing (MACRES) is well equipped with facilities for Geographic Information System (GIS) and remote sensing. It regularly analyses remote sensing data and produces maps, showing information related to land matters for various parts of the country. An example is the land cover map: a series of maps produced at a scale of 1:50,000. The maps, which show all types of land use, are developed from interpretations of satellite imagery. Such maps, when available in a time series, can be used to study land use and vegetation changes, expansion in agricultural land, extent of idle land, reduction in forest cover, conversion of agricultural land to other uses, and so on.

Native people in Malaysia are known as Orang Asli. No land titles are being given to these people, with exception of Johor and Pahang States in 1997, who have acknowledged their rights and handed over land titles to some extent. Many states are expected to follow suit in coordination with Jabatan Hal Ehwal Orang Asli (JHEOA), but the process is painstakingly slow in the past 15 years or so. JHEOA is the governing body of Orang Asli people as aborigines of Malaysia. In accordance to equal right, the reform of Orang Asli land rights must be structural and comprehensive. The Aborigines People Act 1954 does little by way of recognition or conferment of land right; thus, amendments can be made in National Land Code (2002) of Malaysia. Of course, with the consensus of the Orang Asli, the State and the Federal government state that "the special position of the aborigines in respect of land usage and land right shall be recognized."

As a conclusion, soil is a valuable natural resource. In time, human has examined, classified, and utilized soils fundamentally in every aspect of life. Soil has provided mankind with plenty of resources in mineral form and as a base for any structure in any civilization. For example, in mobile phone industry, a mineral commercially known as tantalite (coltan) is highly sought after for electronic capacitor properties, and its application in virtually all electrical appliance one can imagine. Some years ago, tantalum mining was active in Malaysia as by-product of tin mining. This in return influenced some of the landscape that we see today in Malaysia, ex-mining areas that are mostly idle/abandoned land today.

Agronomist, soil scientist, engineers, policy makers, government, and private sectors all look at soil in a different perspective; however, all may agree on one thing that soil sustains life. Malaysia has a good collection of database on soil survey and interpretation of 1960–1980. However, there have been many changes as the country transformed from an agrarian society (from 1960s postindependence) to the present-day industrialized nation. Thus, focus on soil survey, data, and interpretation require new insights. With that, some major issues that can be further updated include the following:

1. Land-use pattern with spatial variability changes
2. Soil erosion and their impact on development
3. Land degradation and their impact on ecosystem
4. Sustainability of arable and nonarable land
5. Optimization of idle/abandoned land to reduce opening new area (forest areas)
6. Micromanagement and training of small-scale farmers
7. Soil ecology impact assessment from forest clearing for timber and/or plantation
8. Land subjected to natural hazard due to changes in topography
9. Risk management (diseases, climate change, and so on)
10. Postdisaster (i.e., recent flood disaster in east coast of Peninsular Malaysia)

Advancement of technology in the last 20 years with hardware and software such as Global Positioning System (GPS), GIS, ArcGIS, MapInfo, and many more is able to improve traditional cartography, with simulation and computer modeling. Fast, neat, and swift decision can be made in the shortest time with the widest information due to these technological advancement. Thus, adaptation of technology is the way forward in sustainable use of soil in the future.

REFERENCES

Aminuddin, B.Y., Chow, W.T. and Ng, T.T. (1990). Resources and problems associated with sustainable development of upland areas in Malaysia. In *Technologies for Sustainable Agriculture on Marginal Uplands in South-east Asia* (Blair, G. and Lefroy, R., Eds.), ACIAR Proceedings no. 33, Australian Centre for International Agricultural Research, Canberra, pp. 55–61.

Azima, A.M. and Ismail, O. (2011). Idle agriculture land resources management and development– An institutional causes: A case study in Malaysia. *Interdisciplinary Journal of Research in Business*, 1(8), 21–27.

Consumer Association Penang (CAP). (2000). *Land Issues in Malaysia*. Consumer' Association of Penang (CAP), Penang, Malaysia.

Department of Agriculture (DOA). (2013). *Garis Panduan: Pembangunan Pertanian Di Tanah Bercerun*. Putrajaya, Malaysia.

DOA-MARDI. (1993). Recommendations of the workshop on the preparation of guidelines for agricultural activities on steepland, July 5–8, 1993, Cameron Highlands.

Ely, R.T. and Wehrwein, G.S. (1940). *Land Economics*. Macmillan Company, New York.

Gurmit, S., Tan, Y.P., Padman, C.V.R. and Lee, L.W. (1986). Experiences on the cultivation and management of oil palms on deep peat in United Plantations Berhad. In: *Proceedings of the 2nd International Soil Management Workshop*, Thailand/Malaysia, April 7–8, 1986.

Hashim, G.M. (2009). Salt-Affected Soils of Malaysia. virualcentre.org (Retrieved: December 10, 2014).

Huang, Y. (1972). Some reflections on padi double-cropping in West Malaysia. *Malayan Economic Review*, 17(1), 119–129.

Smit, J. and Nasr, J. (1992). Urban agriculture for sustainable cities: Using wastes and idle land and water bodies as resources. *Environment and Urbanization*, 4(2), 141–152.

National Land Code. (2002). *A Manual on the National Land Code*. Koperasi Pegawai Pentadbbiran dan Pengurusan Tanah Malaysia Berhad, Kuala Lumpur, Malaysia.

Paramananthan, S. (2014). *Selected Papers on Soil Science Problem Soil*. Agricultural Crop Trust, Petaling Jaya, Malaysia.

Sahak, M. (1986). Memperdagangkan tanah terbiar dalam konteks beberapa polisi kerajaan. *Paper presented at the National Conference on Padi*, Serdang, Selangor, January 20–22, 1986.

Shamshuddin, J. (2006). *Acid Sulfate Soils in Malaysia*. UPM Press, Serdang, Malaysia.

Siong, T.C. (2003). *Classification and utilization of the steep land for Agriculture in Sarawak*. Soil Management Branch, Department of Agriculture, Sarawak.

Index

Note: Page numbers followed by f and t refer to figures and tables, respectively.